Lecture Notes in Mechanical Engineering

Lecture Notes in Mechanical Engineering (LNME) publishes the latest developments in Mechanical Engineering—quickly, informally and with high quality. Original research or contributions reported in proceedings and post-proceedings represents the core of LNME. Volumes published in LNME embrace all aspects, subfields and new challenges of mechanical engineering.

To submit a proposal or request further information, please contact the Springer Editor of your location:

Europe, USA, Africa: Leontina Di Cecco at Leontina.dicecco@springer.com
China: Ella Zhang at ella.zhang@cn.springernature.com
India, Rest of Asia, Australia, New Zealand: Swati Meherishi
at swati.meherishi@springer.com

Topics in the series include:

- Engineering Design
- Machinery and Machine Elements
- Mechanical Structures and Stress Analysis
- Automotive Engineering
- Engine Technology
- Aerospace Technology and Astronautics
- Nanotechnology and Microengineering
- Control, Robotics, Mechatronics
- MEMS
- Theoretical and Applied Mechanics
- Dynamical Systems, Control
- Fluid Mechanics
- Engineering Thermodynamics, Heat and Mass Transfer
- Manufacturing Engineering and Smart Manufacturing
- Precision Engineering, Instrumentation, Measurement
- Materials Engineering
- Tribology and Surface Technology

Indexed by SCOPUS, EI Compendex, and INSPEC.

All books published in the series are evaluated by Web of Science for the Conference Proceedings Citation Index (CPCI).

To submit a proposal for a monograph, please check our Springer Tracts in Mechanical Engineering at https://link.springer.com/bookseries/11693.

Kosmas Alexopoulos · Sotiris Makris ·
Panagiotis Stavropoulos
Editors

Advances in Artificial Intelligence in Manufacturing II

Proceedings of the 2nd European Symposium
on Artificial Intelligence in Manufacturing,
October 16, 2024, Athens, Greece

 Springer

Editors
Kosmas Alexopoulos
Laboratory for Manufacturing Systems
and Automation
University of Patras
Patras, Greece

Sotiris Makris
Laboratory for Manufacturing Systems
and Automation
University of Patras
Patras, Greece

Panagiotis Stavropoulos
Laboratory for Manufacturing Systems
and Automation
University of Patras
Patras, Greece

ISSN 2195-4356   ISSN 2195-4364 (electronic)
Lecture Notes in Mechanical Engineering
ISBN 978-3-031-86488-9   ISBN 978-3-031-86489-6 (eBook)
https://doi.org/10.1007/978-3-031-86489-6

This work was supported by Teaching Factory Competence Center.

This Springer imprint is published by the registered company Springer Nature Switzerland AG
The registered company address is: Gewerbestrasse 11, 6330 Cham, Switzerland

If disposing of this product, please recycle the paper.

Preface

This volume of Lecture Notes in Mechanical Engineering is dedicated to the proceedings of the 2nd European Symposium on Artificial Intelligence in Manufacturing (ESAIM 2024), which was held in Athens, Greece, on 16th of October 2024. The conference was organized by the Teaching Factory Competence Center (TF-CC) with the support of the Laboratory for Manufacturing Systems & Automation (LMS) of the University of Patras, Greece.

Artificial intelligence (AI) has been recognized on a global scale as one of the cornerstone technologies shaping the digital transformation of modern societies and industries. The ESAIM 2024 attempts to establish a continuum in the development of AI from deep learning models to generative AI ones, focusing on the field of manufacturing. However, due to the complexity of systems, equipment, and operations there are several challenges which have not yet been addressed. Concretely, these challenges include among others, data availability and processing, system interoperability, scalability, ethical and regulatory compliance, and high implementation and/or maintenance/update costs. These challenges call for more advanced methods for data standardization and curation, modular design of AI tools and algorithms for expandability and scalability in an affordable manner. Therefore, the scope of ESAIM 2024 spans a plethora of topics around AI development and implementation within modern Industry, which are summarized below:

- Process monitoring, optimization, and control.
- Condition monitoring, diagnosis, and predictive maintenance.
- Quality assessment and prediction.
- Production planning, scheduling, and control of manufacturing systems and value chains.
- Flexible and precise robotics, enhanced human-robot collaboration.
- Design of manufacturing systems, equipment, processes, and products.
- Digital platforms, data spaces, and information technologies for AI applications in manufacturing systems.
- Digital Twin to optimize process, equipment, and plant operations.
- Data augmentation and synthetic data for developing AI applications.
- Education and training for developing skills for AI in manufacturing.
- Ethical and legal aspects of AI in manufacturing.

Considering the interdependence of Industry 4.0 and Industry 5.0 technologies and techniques, ESAIM 2024 actively supports research work on cross-cutting aspects such as information systems, regulation, education, systems engineering, and data augmentation.

The current volume of ESAIM comprises 29 research works from organizations and researchers in Europe. The symposium has welcomed contributions focusing on theoretical, applied research, and industrial case studies. In comparison with the first volume of ESAIM 2023, an increase of approximately 52% of the final accepted papers has been achieved, indicating the developing dynamic of the symposium and of the AIM-NET network. The organizing committee would like to thank all the authors for their valuable contributions. Furthermore, the organizers would like to thank the members of the Program Committee and invited external reviewers for their efforts and expertise in contributing to the paper review process, without which it would be impossible to maintain the high standards of peer-reviewed papers.

The current volume of the book entitled "Advances in Artificial Intelligence in Manufacturing II" consists of five sections. The first section "AI in Process Level" focuses on developments in AI applications for monitoring, optimization, and control of manufacturing processes. Key topics include digital twins for process simulation and optimization, as well as recurrent convolutional neural networks (RNNs) for defect detection, real-time process monitoring for zero-defect manufacturing, anomaly detection in additive manufacturing, development of active learning methods for defect detection, generation of synthetic data, and explainable AI approaches for surrogate modeling.

In Section 2, "AI in Equipment Level", innovative AI applications for improving manufacturing equipment performance and human-machine interaction are investigated. The main focus areas in this section include Generative AI (GenAI) for synthetic data composition in robotics application, development of AI-based assistants for improving manufacturing flexibility and human-machine interaction, federated learning for data curation, and advances in machine vision.

In Section 3, "AI in Systems Level" AI-based optimization of manufacturing systems and value chains is investigated. Papers include AI-assisted scheduling in flexible job shop, agent-based communication for fault diagnosis in skill-based production, utilizing Industry 4.0 language and Asset Administration Shells (AAS), AI-based decision support systems as well as AI-based solutions for mass customization, enabling advanced strategies for meeting diverse customer needs while maintaining efficiency and scalability across manufacturing systems.

Section 4, "Generative AI and Large Language Models (LLM)" focuses on cutting-edge AI technologies in manufacturing. Concretely, topics include integration of LLMs with augmented reality (AR) for industrial maintenance optimization, GenAI-based design of robotic cells in combination with virtual reality (VR), vector-based anomaly detection, automated instruction generation using retrieval-augmented generation (RAG), natural language processing (NLP) for product lifecycle efficiency through materials insight and optimization.

The volume concludes in Section 5, "Fundamental AI Topics", in which the core principles and emerging trends for AI applications in manufacturing are covered. The main topics of this section include frugal AI for affordability and resource optimization, explainable AI challenges identification, federated learning for the development of collaborative and decentralized AI models, and ethical considerations/data governance/privacy management for AI solutions in small and medium enterprises (SMEs), highlighting the importance of responsible and scalable AI integration.

We would like to express our gratitude to EIT Manufacturing and EIT Manufacturing Region South East center for sponsoring the symposium.

Finally, we appreciate the partnership with Springer, EasyChair, and the members of the Artificial Intelligence in Manufacturing Network (AIM-NET) for their essential support during the preparation of ESAIM 2024.

December 2024   Kosmas Alexopoulos
                                                                      Sotiris Makris
                                                            Panagiotis Stavropoulos

Organizing Committee

Chairmen

Kosmas Alexopoulos (Assoc. Prof.)	Laboratory for Manufacturing Systems & Automation (LMS), Greece
Sotiris Makris (Assoc. Prof.)	Laboratory for Manufacturing Systems & Automation (LMS), Greece
Panagiotis Stavropoulos (Assoc. Prof.)	Laboratory for Manufacturing Systems & Automation (LMS), Greece

Organizing Committee

Panagiotis Aivaliotis	Teaching Factory Competence Center, Greece
Konstantina Salagianni	Teaching Factory Competence Center, Greece
Christina Amalia Sotiropoulou	Teaching Factory Competence Center, Greece
Konstantinos Dimoulas	Teaching Factory Competence Center, Greece
Giannis Angelopoulos	Laboratory for Manufacturing Systems & Automation (LMS), Greece
Zoi Arkouli	Laboratory for Manufacturing Systems & Automation (LMS), Greece
Katerina Bakopoulou	Laboratory for Manufacturing Systems & Automation (LMS), Greece
Harry Bikas	Laboratory for Manufacturing Systems & Automation (LMS), Grcccc
Dimosthenis Dimosthenopoulos	Laboratory for Manufacturing Systems & Automation (LMS), Greece
Christos Gkournelos	Laboratory for Manufacturing Systems & Automation (LMS), Greece
Panagiotis Karagiannis	Laboratory for Manufacturing Systems & Automation (LMS), Greece
Apostolis Papavasileiou	Laboratory for Manufacturing Systems & Automation (LMS), Greece

Scientific Committee

Alexopoulos Kosmas	LMS, Greece
Almeida Antonio	INESC TEC, Portugal

Arkouli Zoi	LMS, Greece
Arrieta Juanan	IDEKO, Spain
Bediaga Iñigo	IDEKO, Spain
Brajovic Danilo	Fraunhofer IPA, Germany
Bikas Harry	LMS, Greece
Burget Pavel	CTU, Czech Republic
Cerquitelli Tania	Politecnico di Torino, Italy
Colosimo Bianca Maria	Politecnico di Milano, Italy
Fernandez Gomez Aitor	IDEKO, Spain
Fernández Martínez Andrea	AIMEN, Spain
Gadeyne Klaas	Flanders Make, Belgium
Grandal Montero Santiago	AIMEN, Spain
Grasso Marco Luigi Giuseppe	Politecnico di Milano, Italy
Gusmeroli Sergio	Politecnico di Milano, Italy
Hashemi Petroodi Seyyed Ehsan	IMT-Altantique, France
Helaakoski Heli	VTT, Finland
Hengel Katharina	DFKI, Germany
Huber Marco	Fraunhofer IPA, Germany
Jurmu Marko	VTT, Finland
Kadera Petr	CTU, Czech Republic
Kolář Petr	CTU, Czech Republic
Makris Sotiris	LMS, Greece
Michalos George	LMS, Greece
Meyers Bart	Flanders Make, Belgium
Montero Castro Ignacio	AIMEN, Spain
Muiños Landi Santiago	AIMEN, Spain
Mulder Wico	TNO, The Netherlands
Nikolakis Nikolaos	LMS, Greece
Olsen Mikkel Labori	DTI, Denmark
Orlandini Andrea	CNR, Italy
Outón Méndez José Luis	TECNALIA, Spain
Papacharalampopoulos Alexios	LMS, Greece
Pedrocchi Nicola	CNR, Italy
Penalva Mariluz	TECNALIA, Spain
Picard Francois	DTI, Denmark
Rahman Mostafizur	MTC, United Kingdom
Ricondo Itziar	IDEKO, Spain
Roth Marco	Fraunhofer IPA, Germany
Sidorenko Aleksandr	DFKI, Germany
Stavropoulos Panagiotis	LMS, Greece
Unamuno Gorka	IDEKO, Spain
Usatorre Luis	TECNALIA, Spain

Van Bekkum Michael	TNO, The Netherlands
Van Dijk Wietse	TNO, The Netherlands
Van Rhijn Gu	TNO, The Netherlands
Vidal Vilariño Félix	AIMEN, Spain
Vienne Caroline	CEA, France
Wagner Achim	DFKI, Germany
Wu Xinyang	Fraunhofer IPA, Germany

Contents

AI in Equipment Level

AI in System Level

Generative AI and Large Language Models (LLM)

Fundamental AI Topics

AI in Process Level

Digital Twins of Manufacturing Processes Under Industry 5.0

Alexios Papacharalampopoulos, Kyriakos Sabatakakis, Olga Maria Karagianni, and Panagiotis Stavropoulos[✉]

Laboratory for Manufacturing Systems and Automation (LMS), Mechanical Engineering and Aeronautics Department, University of Patras, 26504 Patras, Greece
pstavr@lms.mech.upatras.gr

Abstract. Digital Twins of manufacturing processes are expected to boost processes' performance and efficiency, since they are able to run optimization in real-time and provide feedback and/or control to them . At the same time, their adaptivity allows for uncertainty handling, through specific parameters estimation or addressing the processes in a robust way. However, it is still pending to investigate how the emerging Industry 5.0 and the relevant viewpoints of the involved technologies will affect the operation of the digital twins. To this end, herein, a multifold and systematic analysis is run to see how Industry 5.0 affects the digital twin concept, taking into account human centricity in terms of inclusion and empowerment. Next sustainability and resilience are also studied under the same context. To achieve this, a hypothetical case study of Additive Manufacturing optimization is taken into consideration, as well as the link to other manufacturing functionalities (as jobs), for the sake of completeness.

Keywords: Digital Twin · Industry 5.0 · Human Centricity · Sustainability · Resilience

1 Introduction

In recent years, manufacturing processes have become increasingly digital. With the help of Industry 4.0 (I4.0), digitalization leads to various modifications in all production processes, increasing ways of communication and collaboration, and adopting new types of technologies such as Digital Twins (DTs) [1]. It is a dynamic digital model, that reproduces the behavior, characteristics and interactions of the physical space in real or near real-time. DTs are created using data collected from sensors, the Internet of Things (IoT) and other sources embedded in the physical environment, leading to a vast amount of data that can be processed with advanced analytics and/or Artificial Intelligence (AI). The results of the analysis can then be used to improve the performance of the process in the physical space [2, 3]. In practical terms a DT can be realized from three perspectives, combining three concepts: multi-layer, multi-level, and multi-perspective First, it is a computer application deployable on various computing systems that interacts with other entities and applications. Second, it replicates the physical part of a Cyber-physical System (CPS). Third, it simulates a physical system using virtual replicas [4].

© The Author(s) 2025
K. Alexopoulos et al. (Eds.): ESAIM 2024, LNME, pp. 3–11, 2025.
https://doi.org/10.1007/978-3-031-86489-6_1

The concept of DTs has gained popularity in various industries, including manufacturing, healthcare, automotive, aerospace and others [5] as they are enabling digital transformation, improving operational efficiency and driving innovation. They offer many benefits such as product development and testing, continuous improvement, optimization and simulation, predictive maintenance, etc. With DTs, operators can remotely monitor and control processes, which are in extreme environments, or even dangerous, providing flexibility and accessibility. They also provide valuable information about being alert to an impending problem or potential failures. This is very useful for industries, that need to make quick data-driven decisions to prevent critical situations and help them continuously improve their operations and predict maintenance to minimize downtime (Fig. 1) [6].

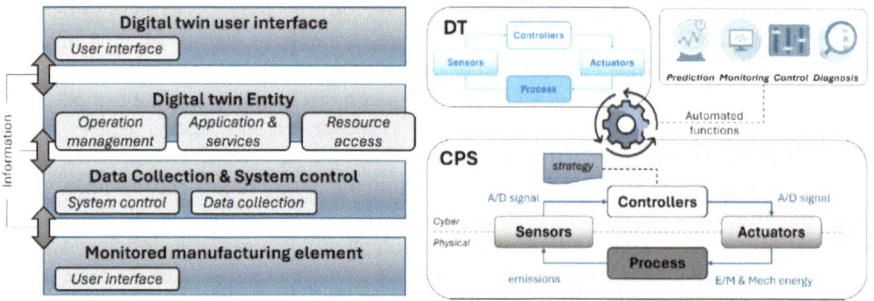

Fig. 1. Digital twin framework based on the description in the ISO 23247 standard [4] (left) The interaction of DTs with the CPS (right)

While automation and digital technologies have revolutionized many aspects of manufacturing, there are several reasons why human involvement remains essential; Industry 5.0 (I5.0) recognizes the importance of human activity in manufacturing. It represents integrating human intelligence with advanced technologies such as artificial intelligence, robotics, IoT and automation. In parallel to I4.0, I5.0 emphasizes collaboration between humans and machines. Machines handle repetitive tasks, data processing, and other routine activities, leaving humans to focus on tasks that require creativity, critical thinking, problem-solving, and emotional intelligence [7]. This collaboration enables more flexible and adaptable manufacturing processes, where humans and machines work together in symbiosis, enhancing productivity, efficiency, and innovation.

The main challenge of I5.0 is to adopt three main pillars in the manufacturing industry. These pillars are human-centricity, sustainability and resilience. Human centricity places basic human needs and interests at the center of the manufacturing process. By doing that, companies can promote an inclusive and fulfilling work environment. This is achieved by motivating employees to upgrade, better career opportunities, and boosting self-confidence and well-being. Providing opportunities for skill development, career development and boosting employee confidence not only benefits individual workers but also contributes to the overall success and innovation of the industry [8].

As a result, industry workers are no longer treated as a plain resource in the production process but as an investment, as valuable assets in which money must be invested to increase their motivation and productivity. This shift in perspective encourages companies to invest in their employees' development and leads to increased job satisfaction and productivity [7]. Besides the change in the human approach, the environmental approach of the industry should also change (Sustainability). I5.0 focuses on environmental protection as a priority, integrating sustainable practices into manufacturing, i.e. reducing waste and optimizing energy use. Also, nowadays people are more aware of the environment, and the sustainability of the planet in terms of increasing waste, excessive use of natural resources, etc. By embracing sustainability, companies can enhance their reputation, attract environmentally conscious consumers and mitigate the risks associated with climate change and resource scarcity. Sustainable manufacturing practices not only benefit the environment but also contribute to long-term profitability and market competitiveness [6, 9, 10]. Resilience refers to the ability of a system to maintain or rapidly recover to a steady state, during and after a major disaster caused by (geo-) political changes, natural emergencies, supply chain issues, or economic downturns. I5.0 emphasizes the importance of creating resilient manufacturing processes that can withstand unexpected challenges and maintain operations with minimal disruption. This may include diversifying supply chains, implementing agile manufacturing systems and investing in technologies such as predictive maintenance and real-time data analytics [9]. By prioritizing resilience, companies can reduce vulnerability to external shocks, improve business continuity and ensure manufacturing stability even in the face of adversity.

Herein, the focus is given to the integration of I5.0 aspects in DTs. The focus is on operations, rather than models, concepts and applications [11, 12]. Hence, with the paradigm of Additive Manufacturing (AM) in mind, the remainder of this paper is structured as follows: Sect. 2 presents a few facts about the main functions of DTs in the manufacturing process. In Sect. 3 the purposes of these are mentioned and are interlinked with I5.0 concepts and finally, in Sect. 4 the conclusions of this work are presented.

2 Background on DTs

As aforementioned, in manufacturing, DTs are expected to be very useful tools. Some of the objectives of a DT in a manufacturing system are to assist decision-making processes, to enable automation of decisions by simulating certain elements and processes within the real system, to use real-time data to run simulation experiments, to consider the different states of the manufacturing system, etc. These goals could be achieved thanks to three main functions of DTs: prediction/diagnosis, monitoring and process control, which are all analyzed below.

The terms prediction & diagnosis, indicate that the actions of the real system are studied before the actual runtime and allow engineers to make informed decisions, optimize operations by testing different parameters, and enhance the reliability and performance of their systems and processes. In more detail with the DTs, engineers can conduct what-if analysis scenarios. They can simulate different operating conditions, environmental factors, or input parameters to predict how the real system will respond in different situations. By predicting how the real system will behave, they can identify potential bottlenecks, inefficiencies, or opportunities for improvement. This predictive insight enables problem identification, and proactive adjustments to optimize performance and improve productivity. It also leads to forecasting of maintenance of machinery, buildings, etc. as well as in choosing the appropriate type of maintenance depending on the situation they have to deal with (reactive, preventive, predictive, regulatory, etc.). Finally, the DT can use online measurements to estimate the fatigue damage to calculate the remaining useful life of the system and plan the corresponding actions accordingly [13–15].

DTs can also help predict and diagnose potential hazards, failures, or damage to products before they occur in the physical system. By analyzing historical data and running predictive algorithms, engineers can predict and prevent problems, thereby reducing downtime, wasting raw materials minimizing component rejection rate and minimizing costly outages. They can also help prevent damage to factory machinery and infrastructure, preventing potential workplace accidents [13, 16].

Monitoring is the collection of live position data from the system processing the simulation. Specifically, DTs continuously receive streams of data from sensors, IoT devices, and other sources connected to the physical system they represent. This real-time data integration enables monitoring of current system status and performance, enabling proactive decision-making, maintenance planning, optimization of operations, etc. DTs also provide a visual representation of the state of the physical system, often through table tools or graphical interfaces. This visualization allows operators and stakeholders to monitor key metrics, trends and anomalies in real-time and receive early intervention to address issues or prevent potential problems before they escalate [12, 16].

DTs allow remote monitoring of physical assets or processes from anywhere with an Internet connection. With friendly-to-use and understandable interfaces, engineers are more likely to remotely monitor systems, and can better use their time to improve synergies and collaborations that lead to greater productivity. This capability is especially valuable for distributed systems or assets located in remote or inaccessible locations [14, 17]. In addition, proactive data monitoring can identify trends and deviations from the normal behavior of a situation or process and study "why" it occurred or "how" to avoid it in the future. By monitoring the data in real-time and comparing it to historical data stored in the DTs, engineers can be guided to identify anomalies in processes and understand the root causes of performance fluctuations. The performance of assets or processes over time can also be monitored. By analyzing historical data and trends, they can evaluate performance metrics, identify areas for improvement, and optimize operations [13, 14]. Analyzing and understanding the unexpected failures or data obtained during monitoring after the actual system operation is called diagnosis. More specifically, when an anomaly or failure is detected, DTs facilitate root cause analysis by providing comprehensive data about system behavior prior to the event. Historical data stored in the DTs allows stakeholders to trace the sequence of events and identify underlying factors contributing to the issue. In this way, failures can be prevented when the factors driving them are observed before they are finalized. By observing data analytics from a process, the interested party can notice or identify potential equipment failures or degradation before they occur. In this way, organizations can proactively schedule maintenance activities, minimizing downtime and optimizing the life of production machinery. In addition, through data visualization, analysis, and simulation capabilities, DTs help operators and engineers make informed decisions for effective problem-solving, preventive maintenance, and more as DTs, provide decision support tools to diagnose problems and recommend appropriate actions or interventions [10]. If failure is not detected early, by analyzing historical data and performance trends, engineers can iterate and improve the design, operation and maintenance of their systems over time. That is, diagnosis with DTs enables stakeholders to learn from past failures or issues and implement corrective actions to avoid recurrence in the future. Finally, another important factor that enhances the robustness of the production process is control. Using DTs for process control can greatly improve the ability to guarantee the desired performance, even in the presence of uncertainties and disturbances [18]. This leads to adaptivity to changes, and process dynamics by continuously learning and updating the DTs. This is crucial for maintaining performance despite a variety of uncertainties/disturbances. This not only improves process reliability and efficiency, but also enhances the ability to anticipate and respond to potential issues proactively.

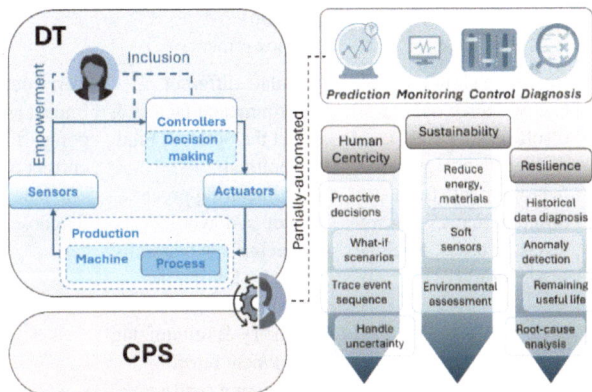

Fig. 2. DT transformation through I5.0 pillars

3   Implications of Industry 5.0 and Applications

The three main functions of DTs play a critical role in I5.0 by enabling predictive capabilities, real-time monitoring, and process control (Table 1), as part of jobs, with AM case study in mind. They enhance and improve production with the capabilities they offer using the respective tools. They also empower the employees by offering immediately applicable knowledge that helps develop their skills and enhance their interest in their work. Thus, they serve as key enablers for the human-machine collaboration paradigm, assisting manufacturers to achieve greater efficiency, sustainability, and resilience.

DTs provide a powerful framework for two-way information flow between the digital model and the human interacting with it (Fig. 2). This interaction is key to improving decision-making, optimizing operations, and improving overall system performance. User insights ensure that the DT remains relevant and accurately reflects the physical

Table 1. Parameters of DTs according to the pillars of I5.0

	Monitoring	Prediction/Diagnosis	Control
Human Centricity	Monitoring of system status, enabling proactive decision-making, (empowerment)	Simulate what-if scenarios & human chooses the best (inclusion) [19], Early intervention to address issues or prevent potential problems (inclusion & empowerment), Trace event sequences & identify underlying factors that contribute to each event: people can prevent failures or problems (empowerment)	Decision making (empowerment), Trace the sequence of events & identify underlying factors that contribute to each event so that people can prevent them. (empowerment), Handle uncertainty (empowerment)
Sustainability	Real-time simulation of energy sensor outcomes ("soft sensors" [20]), Aggregate energy measurements from different manufacturing levels (process, machine, production) [21]	Simulate different environmental factors & select the optimal, Find the optimal manufacturing process to reduce downtime Reducing raw materials' waste by finding the optimal product geometry & minimizing component rejection, Minimizing costly outages	Carry out environmental impact assessments by comparing historical & current data, Resource efficiency can be diagnosed
Resilience	Remote monitoring of processes in hard-to-reach areas in real-time, Anomaly detection in procedures & their immediate resolution [22], Understand the root causes of performance fluctuations by comparing historical with present data, Identify areas for improvement & optimize operations	Finding the optimal manufacturing process to improve productivity, Proactive settings for optimizing performance & choosing the appropriate type of maintenance, Early assessment of fatigue damage, Remaining useful life calculation and forecasting its replacement of systems	By comparing historical data and current data, damage to products, machinery, etc. can be diagnosed (Human-in-the-Loop), Support decision-making in case of people changing positions

entity's unique circumstances and requirements depending on the scenario or requirements. By facilitating a dynamic exchange of information along with human assistance, DTs bridge the gap between the physical and digital worlds.

I5.0 is understood as an evolution and logical continuation of the existing I4.0 paradigm. Thus, the impact of this "transition" on AI is the adoption of existing methods such as Active & Transfer Learning to implement Human-in-loop Machine Learning (ML) and methods such as Explainable & Frugal AI. These aim to integrate human intelligence to reduce development costs and boost the performance of AI, enhance the trust of stakeholders, and reduce the ever-increasing energy demand of the entire AI development and maintenance value chain.

4   Conclusions and Future Outlook

With the adoption of I5.0, the interaction between DTs and human actors is expected to be immense and provide high potential for productivity and sustainability. However, it seems that the strategies imposed by the I5.0 pillars are descriptive and may be lacking methodologies that will facilitate their integration. Then, it will be easier to prove the superiority of such approaches through manufacturing and monetary KPIs, especially in the case of AM where there is room for further digitalization and job transition.

Future research should develop standardized methodologies for DT integration within I5.0, identify and validate KPIs for quantifying benefits, and explore advanced AI and ML for predictive maintenance and process optimization. Interdisciplinary approaches combining data science, human factors, and environmental science should be considered. Research should also investigate the scalability and flexibility of DT solutions for various manufacturing environments, including SMEs, and examine the ethical and social implications of increased automation and human-machine collaboration to ensure responsible, sustainable advancements.

Acknowledgment. The current work has been supported by EU project BRIDGES 5.0 (101069651).

References

1. Cinar, Z.M., Nuhu, A.A., Zeeshan, Q., Korhan, O.: Digital Twins for industry 4.0: a review. In: Calisir, F., Korhan, O. (eds.) Industrial Engineering in the Digital Disruption Era: Selected papers from the Global Joint Conference on Industrial Engineering and Its Application Areas, GJCIE 2019, pp. 193–203. Springer International Publishing (2020)
2. Tao, F., Zhang, H., Liu, A., Nee, A.Y.C.: Digital Twin in Industry: state-of-the-art. IEEE Trans. Ind. Inform. **15**(4), 2405–2415 (2019)
3. Pérez, L., Rodríguez-Jiménez, S., Rodríguez, N., Usamentiaga, R., García, F.D.: Digital Twin and virtual reality based methodology for multi-robot manufacturing cell commissioning. Appl. Sci. **10**(10), 3633 (2020)
4. Mohammed, W.M., Haber, R.E., Martinez Lastra, J.L.: Ontology-driven guidelines for architecting digital twins in factory automation applications. Machines **10**(10), 861 (2022)

5. Attaran, M., Celik, B.G.: Digital Twin: benefits, use cases, challenges, and opportunities. Decis. Analyt. J. **6**, 100165 (2023)

6. Sharma, A., Kosasih, E., Zhang, J., Brintrup, A., Calinescu, A.: Digital Twins: state of the art theory and practice, challenges, and open research questions. J. Ind. Inf. Integr. **30**, 100383 (2022)

7. Aheleroff, S., Huang, H., Xu X., Zhong, R.Y.: Toward sustainability and resilience with Industry 4.0 and Industry 5.0. Front. Manuf. Technol. **2**, 951643 (2022)

8. Xu, X., Lu, Y., Vogel-Heuser, B., Wang, L.: Industry 4.0 and Industry 5.0 – Inception, conception and perception. J. Manuf. Syst. **61**, 530–535 (2021)

9. Mourtzis, D.: Digital twin inception in the Era of industrial metaverse. Front. Manuf. Technol. **3**, 1155735 (2023)

10. Leng, J., Sha, W., Wang, B., Zheng, P., et al: Industry 5.0: prospect and retrospect. J. Manuf. Syst. **65**, 279–295 (2022)

11. Lv, Z.: Digital twins in industry 5.0. Research **6**, 0071 (2023)

12. Wang, B., Zhou, H., Li, X., Yang, G., et al.: Human Digital Twin in the context of Industry 5.0. Robot. Comput. Integr. Manuf. **85**, 102626 (2024)

13. Glatt, M., Sinnwell, C., Yi, L., Donohoe, S., Ravani, B., Aurich, C.J.: Modeling and implementation of a digital twin of material flows based on physics simulation. J. Manuf. Syst. **58**, 231–245 (2021)

14. Segovia, M., Garcia-Alfaro, J.: Design, modeling and implementation of Digital Twins. Sensors **22**(14), 5396 (2022)

15. Hasidi, O., Abdelwahed, E.H., Qazdar, A., Boulaamail, A., et al.: Digital Twins-based smart monitoring and optimisation of mineral processing industry. In: 4th International Conference on Smart Applications and Data Analysis, pp. 411–424. Springer International Publishing (2023)

16. Brockhoff, T., Heithoff, M., Koren, I., Michael J., Pfeiffer, J., Rumpe, B.: Process prediction with Digital Twins. In: IEEE International Conference on Model Driven Engineering Languages and Systems Companion, pp. 182–187. IEEE (2021)

17. Verdugo-Cedeño, M., Jaiswal, S., et al.: Simulation-based Digital Twins enabling smart services for machine operations: an industry 5.0 approach. Int. J. Hum. Comput. Interact. 1–17 (2023)

18. Stavropoulos, P., Papacharalampopoulos, A., Michail, K.C., Chryssolouris, G.: Robust additive manufacturing performance through a control oriented Digital Twin. Metals **11**(5), 708 (2021)

19. Vaneker, T., Bernard, A., Moroni, G., Gibson, I., Zhang, Y.: Design for additive manufacturing: framework and methodology. CIRP Ann. **69**(2), 578–599 (2020)

20. Jiang, Y., Yin, S., Dong, J., Kaynak, O.: A review on soft sensors for monitoring, control, and optimization of industrial processes. IEEE Sens. J. **21**(11), 12868–12881 (2021)

21. Stavropoulos, P., Panagiotopoulou, V.C.: Carbon footprint of manufacturing processes: conventional vs. non-conventional. Processes **10**(9), 1858 (2022)

22. Liu, Z., et al.: Additive manufacturing of metals: Microstructure evolution and multistage control. J. Mater. Sci. Technol. **100**, 224–236 (2022)

Recurrent Convolutional Neural Network Based Defect Detection in Submerged Arc Welding Processes

Estefanía Alexandra Tapia Suárez$^{(\boxtimes)}$, Inés Pérez Couñago, Christian Eike Precker, Juan Manuel Montenegro Fernández, and Santiago Muiños-Landín

AIMEN Technology Centre, 36410 Porriño, Spain
estefania.suarez@aimen.es

Abstract. Detecting defects predictively during the welding process, such as porosity, is of vital importance as it allows for the avoidance of degradation in the quality, durability, and productivity of the weld. Research into predictively identifying these types of defects in Submerged Arc Welding (SAW) is quite limited due to the difficulty of gathering data along the process. This remains a challenge to drive the optimization of the manufacturing of pieces that include such welds as the case of large components like pipes in the oil and gas industry. Therefore, this work addresses this challenge and proposes a methodology based on a deep hybrid neural network called recurrent convolutional neural network (RCNN). This deep learning model is capable of detecting and predicting surface porosity defects in real-time using the continuous voltage electrical signal from the SAW process. The training of the RCNN model involved using various weld beads, some with surface porosity and others without. On the one hand, defects were labeled based on the location of the pores along the weld, while on the other hand, the voltage electrical signals were processed and organized. The proposed framework based on RCNN was tested in other weld beads, where the results were satisfactory with the model achieving a high accuracy rate of around 80% in predictive pore detection. Moreover, the model's processing time was <10 ms, meeting the requirements for real-time applications.

Keywords: Submerged arc welding · Porosity · Weld quality · Predictive detection · Deep learning

1 Introduction

Submerged Arc Welding (SAW) is a fusion welding process widely utilized across various sectors, such as oil and gas and shipbuilding. Due to its important applications, quality standards for components manufactured via this process are becoming increasingly stringent. Nonetheless, the emergence of defects, notably porosity, poses a significant challenge in SAW. Such defects result in substantial strength deterioration and alterations in the mechanical properties of the weld, consequently diminishing the durability and productivity of manufactured parts [1]. While ultrasonic inspection and

© The Author(s) 2025
K. Alexopoulos et al. (Eds.): ESAIM 2024, LNME, pp. 12–20, 2025.
https://doi.org/10.1007/978-3-031-86489-6_2

radiographic inspection have traditionally served as nondestructive testing methods for defect detection, their utility is constrained by limitations in real-time detection, high costs, and applicability in large-scale production lines [2].

Different research studies have addressed this issue by developing online monitoring systems that utilize information from the welding process, including voltage, current, temperature, acoustic signals, images, etc.; information closely linked to weld quality [3, 4]. However, in the case of SAW, the use of process parameters for developing online monitoring and defect detection algorithms has been limited, primarily due to challenges related to data collection and handling, as well as the complexity of the process. Relevant developments have been observed only in [5, 6]. In these studies, learning models were implemented using images to detect online defects and monitor bead geometry, respectively. However, utilizing images requires special sensors to capture them due to the obscured view of the weld pool and the presence of flux. Other sensors, such as those for capturing electrical signals, offer greater advantages. These devices simplify real-time signal acquisition and are less susceptible to interference from the complex welding environment, providing greater stability. Specifically, the voltage signal has shown a strong correlation with the physics of the welding process [7]. Consequently, many studies in other processes such as GMAW or TIG have utilized this signal for online defect detection [8–10].

Deep learning (DL) techniques are currently a field of exponential evolution under scientific research and industry. Researchers have deployed DL techniques in online quality inspection to detect potential defects in various welding processes [11]. Convolutional neural networks (CNNs) are among the most representative deep learning algorithms and are widely used in both image-based [5, 12, 13] and sequential signal-based defect classification [14]. However, the CNN model does not consider the temporal connection of the collected signals. Long short-term memory (LSTM), a subset of recurrent neural networks (RNN), is recognized for its prowess in time series analysis and intricate temporal pattern extraction [15]. The LSTM model is characterized by its adaptability and ability to address sequential data, making it particularly suitable for the analysis of electrical measurements of welding processes. The hybrid recurrent convolutional neural network (RCNN) model combines the strengths of both CNN and LSTM, where 1D-CNN extracts relevant special features from the input data and the LSTM layers capture temporal dependencies. RCNN has been successfully applied in various domains such as manufacturing related to welding defects automatic detection [16].

In this context, this work proposes a methodology for the real-time detection of porosity-related defects in SAW process. This methodology utilizes voltage signals in conjunction with the implementation of the deep hybrid neural network RCNN. The RCNN model offers several advantages, including automatic capturing of spatial and temporal features of the data, as well as high potential for applicability and performance. Consequently, the proposed methodology is capable of predicting porosity occurrences based on the real-time progress of the weld. This serves as the foundation for the development of an adaptive control strategy, ultimately leading to the advancement of intelligent welding techniques.

2   Experimental System

The schematic diagram representing the SAW experiment system is depicted in Fig. 1. This diagram comprises the welding machine (Miller Summit Arc 1250), the control module, and the measurement sensors. The sensors utilized in this study, as shown in the diagram, are electrical sensors consisting of differential voltage and current probes. Specifically, the differential probes were installed between the welding machine and the welding plate. Synchronized voltage and current signals are recorded using an oscilloscope (PicoScope 2000) at a sampling rate of 2500 samples per second. The data is swiftly transferred to a computer for analysis and/or application development via a high-speed USB connection.

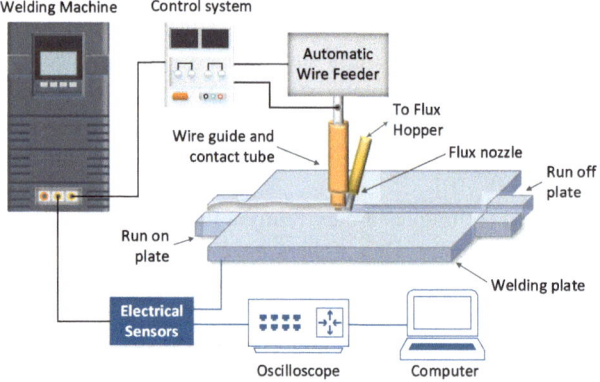

Fig. 1. Schematic diagram of SAW experimental system.

Several experiments were conducted on steel plates of various lengths, using the filler material 3.2 mm OE-SD3 and OP121TT flux. Additionally, different welding parameter settings were employed, as summarized in Table 1.

Table 1. Normal welding settings parameters of SAW

Voltage	Current	Wire Feed Speed	Travel Speed
27–30 V	450–675 A	1.6–2.7 m/min	48–80 cm/min

To induce pore occurrence artificially, welding was performed in some experiments with plates and flux under humid conditions. Authorized personnel visually inspected the surface beads for defects along their entire length and recorded pores of various sizes, ranging from 1 to 2.4 mm in diameter, that had formed along each bead. Similarly, experiments were conducted under ideal conditions where no defects occurred. This was done to evaluate the RCNN model's capability to detect pores in beads with and without porosity.

3   RCNN-Based Porosity Detection

3.1   Model Overview

The hybrid deep learning RCNN model, composed of the CNN and LSTM models, can establish a relationship between voltage measurements of the SAW process and the porosity state. In this way, when the model is applied in real time during the welding process, there will be predictive information on the presence or absence of pores, which represents the basis for implementing control actions. Figure 2 shows the overall framework of the RCNN model.

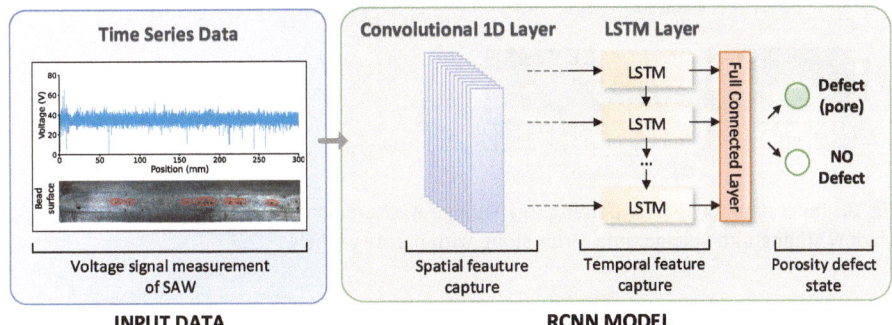

Fig. 2. Overall framework of RCNN model.

As shown in Fig. 2, voltage measurements obtained during the welding process are organized into sequential data and input into the RCNN model. The 1D-CNN layer is responsible for detecting local features and capturing short-term relationships from the electrical signal, such as trends or fluctuations that may not be seen at a glance. This enables the LSTM layer to be more efficient in capturing long-term temporal dependencies throughout the electrical signal sequence, thereby strengthening the anomaly detection capacity. This means CNN and LSTM complement each other in applications like this work, which focuses on porosity detection. Then, the fully connected layers establish relationships among the captured features, and the classifier provides the final determination, indicating the presence or absence of porosity defects.

3.2   RCNN Training

The input data for training the RCNN model consists of voltage measurements obtained from various weld beads and corresponding porosity state labels. Voltage measurements undergo a process known as window sliding, wherein the data is partitioned into time series with a specified window length (measured in millimeters of welding). Conversely, porosity state labeling is conducted based on the visual inspection record of the weld beads.

Figure 3(a) displays the voltage signal obtained during a welding bead with various pores. Figure 3(b) represents the porosity state labeling of this bead for each weld millimeter, where '0' indicates the pores' absence and '1' indicates the pores' presence. Finally, Fig. 3(c) illustrates the RCNN input data, comprising a voltage time series of a 4 mm length (depicted by the green shadow), along with the corresponding porosity labeling in the subsequent millimeter (depicted by the orange shadow). Consequently, each time series is trained to predict the porosity state in the following weld millimeter.

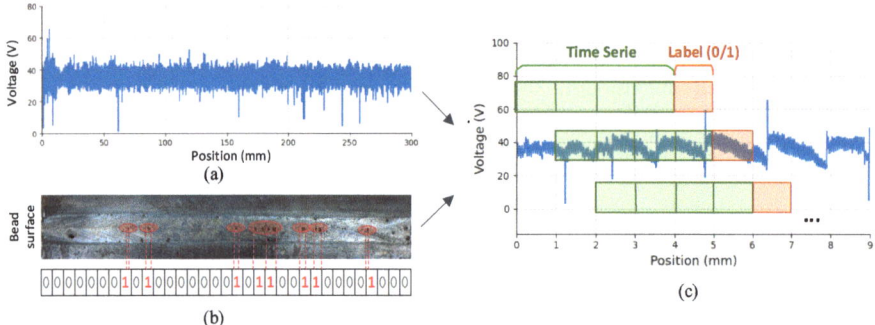

Fig. 3. Input data for RCNN training. (a) Voltage measurement; (b) Porosity state labeling; (c) window sliding into voltage time series along with porosity labels.

RCNN training involves establishing its architecture along with selecting hyper-parameters that minimize the disparity between the model predictions and the true classification states. The established architecture of the RCNN is depicted in Fig. 4.

The convolutional layer includes specifications such as the number and size of convolutional kernels, while the LSTM layer comprises the number of memory units. Additionally, the classifier is composed of fully connected layers and corresponding activation functions, which determine the class with the highest probability.

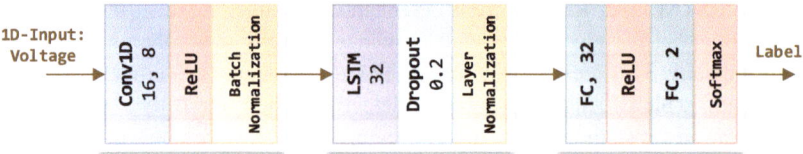

Fig. 4. Representation of the RCNN architecture for porosity state classification.

It is worth mentioning that the loss function used is Weighted Cross-Entropy (WCE) [17]. This function has the advantage of addressing the class imbalance present in this classification task, where the minority class—porosity presence—is of greater interest. By incorporating a weight factor α, WCE allows for establishing a balance between the two classes. Finally, in addition to the training model, an early stopping criterion was employed to avoid overfitting. This criterion involved monitoring the loss of the model on the test data. If the loss did not improve over 10 consecutive epochs, the training process was stopped. The model weights were then restored to the epoch where the best metric was achieved until the end of the training.

4   Results and Discussions

The RCNN deep learning model was developed in Python version 3.10.12 using TensorFlow 2.15.0 on a computer with an Intel Core i7-13800H @ 2.5 GHz processor and 32 GB of RAM. The training process was accelerated by utilizing a GPU machine, which allowed for linear scaling of the mini-batch size and faster computations.

The training was performed using data from 6 surface beads with porosity along their length. The six beads, ranging in length from 300 to 500 mm, make up a total length of 2050 mm. The time series window length was adjusted based on two premises: a) the model achieves great performance in the classification task, and b) the time required by the model to predict the porosity state is as short as possible, allowing sufficient time to implement some control action. In this context, RCNN training was performed with different window sizes of 1 mm, 2 mm, 3 mm, 4 mm, and 5 mm. The final window size will be the one that best meets the two mentioned premises.

Additionally, the database needed to undergo a normalization process due to the different voltage ranges handled in the welding beads. This process was carried out using the z-score algorithm [18]. Once the database was prepared in terms of its partitioning, window size, and normalization, RCNN training was conducted using the architecture mentioned in Sect. 3.2. The classification performance on the test dataset, as well as the RCNN execution time, is summarized in Table 2.

Table 2. RCNN performance and time execution results (test data)

Window size	Precision	Recall	Time execution
1 mm	66,7%	71,6%	6,2 ms
2 mm	69,3%	76,4%	6,7 ms
3 mm	71,2%	76,6%	10,8 ms
4 mm	71,5%	78,3%	10,9 ms
5 mm	78,8%	80,9%	14,2 ms

From Table 2, the ability of RCNN to detect pores improves as the input window size increases. The best performance is achieved with an input window of 5 mm, where precision and recall are around 80%. However, as the input window increases, the time required by RCNN to detect the porosity state (PS) also increases. Given the real-time application requirements of this development, the PS detection time must be very fast (before porosity develops). Considering the fastest welding speed of Table 1 (80 cm/min), the time required to weld 1 mm is 75 ms, which implies that the detection time for PS should be less than 7.5 ms (10% of the time to weld 1 mm) to meet real-time application requirements. From Table 2, it is observed that with a window size of 2 mm, the PS detection time is 6.7 ms, meeting the real-time application requirements, however, the model's performance is reduced to around 70% with this window size.

Given that in a real scenario, pores along the weld are the minority meaning that when one occurs, the RCNN model must be able to identify it, this work prioritizes the recall metric since it reflects the model's ability to identify minority instances (pores). Therefore, with a window size of 2 mm, where the PS detection time is only 6.7 ms, although the precision declines, the recall remains at a competitive value of 76.4%. Thus, this window size is selected for real-time applications.

In this case, the database size is 2420, which was generated based on a window size of 2 mm, a window slide of 1 mm, a sampling frequency of 2500 Hz, and a welding travel speed in the range of 48 to 80 cm/min. The data were partitioned into training (80%) and testing (20%), with the test data corresponding to different chunks of weld beads. Figure 5 shows the confusion matrix of the model on the test data, and Fig. 6 exhibits the model's behavior on a chunk of weld, where a binary indicator represents the detection.

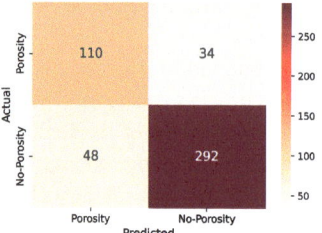

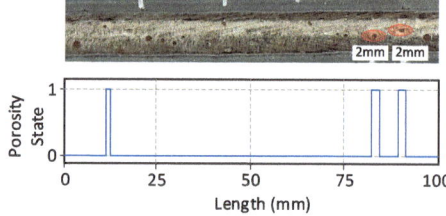

Fig. 5. Confusion matrix results on the test data.

Fig. 6. Porosity detection of RCNN model on a test weld chunk.

The confusion matrix in Fig. 5 shows the model's classification results on the test data. As can be seen, the model performs adequately in porosity detection, with a low presence of false positives (incorrect detection of porosity where it is not present) and false negatives (incorrect detection of porosity where it is present). Although these values are low, they could be improved. For example, increasing the database, especially with more instances containing pores, could provide a better balance between the classes, reduce the number of false positives and false negatives, and thus improve RCNN's performance.

The welding chunk shown in Fig. 6 has two pores in its final part, each with a diameter of 2 mm, which have been highlighted in red. Although a greater number of pores appear to be present throughout the welding, these are actually surface-coating oxidations that do not form cavities or holes like pores do. As shown in Fig. 6 through the binary indicator, the existing pores are detected, with only one false positive in this weld small chunk. Therefore, effective detection of pores, even of different sizes, is achieved through the application of the RCNN model, demonstrating its potential to meet real-time predictive requirements. Improving the classification results of the RCNN could also be achieved by adding another input variable to the model that is related to the defect under study. For example, the variable related to welding plate temperature would be a suitable candidate, as it reflects the plate's heating state and the possibility of trapped gases, which can lead to the generation of porosity.

5 Conclusions and Future Implications

This work proposed online porosity defect detection in the SAW process. The feasibility of this approach was assessed based on the real-time classification performance of the hybrid RCNN model. The results showed the model's significant ability to detect pores throughout the welding execution, with a recall score of 76.4% and an inference time <7.5 ms per millimeter of welding, i.e., less than 10% of the time required to weld 1 mm.

Furthermore, the classification results demonstrated that the variable related to voltage has a close relationship with pore-related defects. As future research to improve the RCNN classification results, the addition of other variables that may relate to the defect under study will be proposed. For example, the welding plate temperature could be a variable that enhances pore detection, given that it reflects the plate's heating state and the possibility of gases being trapped during pore generation. Additionally, another strategy to improve RCNN performance is to use a greater number of welded beads with porosity along their length. This would allow the model to learn from a wider variety of instances with pores, improving its balance between classes and generalization against previously unseen pores. This can be attributed to various tests aimed at expanding the database with instances of porosity, coupled with an increasingly robust model architecture, resulting in an upward trend in performance metrics.

Acknowledgements. This research has received funding from the European Union's Horizon 2020 research and innovation program under the project PENELOPE with Grant Agreement 958303, as well as from the European Union's Horizon 2021 program under the project ZDZW with Gran Agreement 101057404. The authors want to thank the comments and fruitful discussions with all the members of the Smart Systems and Smart Manufacturing (S3M) Unit of the AIMEN Technology Centre.

References

1. Shin, S., Rhee, S.: Porosity characteristics and effect on tensile shear strength of high-strength galvanized steel sheets after the gas metal arc welding process. Metals (Basel) **8**, 1077 (2018)
2. Barot, R.S., Patel, V.J.: Process monitoring and internet of things feasibility for sub-merged arc welding: state of art. Mater. Today Proc. **45**, 4441–4446 (2021)
3. Huang, Y., Xu, S., Yang, L., Zhao, S., Shi, Y.: Defect detection during laser welding using electrical signals and high-speed photography. J. Mater. Process. Technol. **271**, 394–403 (2019)
4. Ma, D., Jiang, P., Shu, L., Gong, Z., Wang, Y., Geng, S.: Online porosity prediction in laser welding of aluminum alloys based on a multi-fidelity deep learning framework. J. Intell. Manuf. **35**, 55–73 (2024)
5. Stavropoulos, P., Papacharalampopoulos, A., Sabatakakis, K.: Online quality inspection approach for Submerged Arc Welding (SAW) by utilizing IR-RGB multimodal monitoring and deep learning. In: Kim, K.-Y., Monplaisir, L., Rickli, J. (eds.) Flexible Automation and Intelligent Manufacturing: The Human-Data-Technology Nexus, pp. 160–169. Springer International Publishing, Cham (2023)
6. FarhatullaBaig, M., Adamsab, K., Dubey, D.: Analysis of Submerged Arc Welding (SAW) surface defects using Convolutional Neural Network (CNN). In: Kumar, H., Jain, P.K., Goel, S. (eds.) Recent Advances in Intelligent Manufacturing, pp. 283–291. Springer Nature Singapore, Singapore (2023)

7. Zhang, Z., Chen, X., Chen, H., Zhong, J., Chen, S.: Online welding quality monitoring based on feature extraction of arc voltage signal. Int. J. Adv. Manuf. Technol. **70**, 1661–1671 (2014)
8. Shin, S., Jin, C., Yu, J., Rhee, S.: Real-time detection of weld defects for automated welding process base on deep neural network. Metals (Basel) **10**, 389 (2020)
9. Li, Y., et al.: A defect detection system for wire arc additive manufacturing using incremental learning. J. Ind. Inf. Integr. **27**, 100291 (2022)
10. Reisch, R.T., et al.: Con-text awareness in process monitoring of additive manufacturing using a digital twin. Int. J. Adv. Manuf. Technol. 1–18 (2022)
11. Papavasileiou, A., Michalos, G., Makris, S.: Quality control in manufacturing – review and challenges on robotic applications. Int. J. Comput. Integr. Manuf. 1–37 (n.d.). https://doi.org/10.1080/0951192X.2024.2314789
12. Zhang, B., Hong, K.-M., Shin, Y.C.: Deep-learning-based porosity monitoring of laser welding process. Manuf. Lett. **23**, 62–66 (2020)
13. Zhang, Z., Wen, G., Chen, S.: Weld image deep learning-based on-line defects detection using convolutional neural networks for Al alloy in robotic arc welding. J. Manuf. Process. **45**, 208–216 (2019). https://doi.org/10.1016/j.jmapro.2019.06.023
14. Su, C., et al.: Improved damage localization and quantification of CFRP using Lamb waves and convolution neural network. IEEE Sens. J. **19**, 5784–5791 (2019)
15. Alsumaidaee, Y.A.M., et al.: Fault detection for medium voltage switchgear using a deep learning hybrid 1D-CNN-LSTM model. IEEE Access (2023)
16. Shang, L., et al.: Deep learning enriched automation in damage detection for sustainable operation in pipelines with welding defects under varying embedment conditions. Computation **11**, 218 (2023)
17. Aurelio, Y.S., De Almeida, G.M., de Castro, C.L., Braga, A.P.: Learning from imbalanced data sets with weighted cross-entropy function. Neural Process. Lett. **50**, 1937–1949 (2019)
18. Singh, D., Singh, B.: Investigating the impact of data normalization on classification performance. Appl. Soft Comput. **97**, 105524 (2020)

On the Use of Generative AI to Support In-Line Process Monitoring in Zero-Defect Manufacturing

Marco Grasso[(✉)], Matteo Bugatti, and Bianca Maria Colosimo

Dipartimento di Meccanica, Politecnico di Milano, Via La Masa 1, 20156 Milano, Italy
marcoluigi.grasso@polimi.it

Abstract. In recent years, the integration of new Artificial Intelligence (AI) tech-niques and capabilities has emerged as one of most promising research fields to aid the industrial development of smart and zero-defect manufacturing solutions. This study explores the potential of generative AI in this field and reviews novel opportunities enabled by generative AI methods, and Generative Adversarial Net-works (GANs) in particular, to aid the generation of augmented datasets including realistic representations of anomalous process patterns. The result is an effective AI framework to learn specific defect features from real data, and reproduce them in an extended way, leading to synthetic but realistic image data that could be used to enhance defect detection and classification performances. The paper reviews the benefits and open challenges associated with the implementation of these techniques, including state-of-the-art examples and real case studies in Additive Manufacturing.

Keywords: generative networks · artificial intelligence · generative adversarial networks · additive manufacturing · image data

1 Introduction

According to Scopus, slightly less than 30k scientific papers devoted to machine learning and artificial intelligence (AI) applications in manufacturing have been published so far, with a rate of more than 1000 per year since 2019. The wide interest for these methods primarily lays in their capacity to solve complex data-driven problems in applications where other statistical techniques exhibit major limitations or can be hardly applicable. The AI market volume in the manufacturing sector is expected to grow from 890 M\$ in 2018 to 21.3 B\$ in 2028 [1]. However, the intrinsic limitations of AI methods, e.g., their black box nature, the need for massive training data and computational resources, still represent a barrier for a widespread industrial adoption in manufacturing applications.

Novel opportunities have been recently triggered by a novel family of techniques known as *generative AI*, which may unlock the potential to tackle and solve a broad range of new problems as well as to revolutionize the way in which traditional problems are addressed [2]. Generative AI consists of a class of machine learning algorithms

© The Author(s) 2025
K. Alexopoulos et al. (Eds.): ESAIM 2024, LNME, pp. 21–30, 2025.
https://doi.org/10.1007/978-3-031-86489-6_3

specifically designed to model the underlying distribution of a dataset aiming to generate new, synthetic instances that resemble the input [2, 3]. This class of methods can capture the statistical properties of the input data in a variety of formats, e.g., text, signals and images, and replicate them.

In the field of smart and zero-defect manufacturing generative models can be adopted to solve different problems, from the generative design of new product concepts to quality assurance and process optimization. This study specifically focuses on synthetic data generation to aid and augment process monitoring and classification performances. The underlying motivation is the following. Real manufacturing data are commonly highly unbalanced, i.e., data collected under normal, stable, and defect-free conditions are commonly more widely available than data gathered in the presence of any actual departure from this state. This represents a limitation for automated anomaly detection and classification, especially when supervised methods are used. To some extents, it also represents a limitation in the field of statistical process monitoring and "one-class-classification" methods [4]: although they rely only on data samples representative of non-defective (in-control) conditions, defective (out-of-control) items and samples are needed to test, validate, and possibly tune their anomaly detection performances. A quite common practice in the literature consists of simulating process data in different out-of-control classes and at different anomaly severity levels. Simulation studies are typically carried out by drawing random samples from pre-defined probability distributions and testing proposed methods in the presence of representative shifts applied to the parameters of the original distribution. However, the higher the complexity of data patterns and formats (e.g., 1D profile data, 2D surfaces, 3D point clouds, images, videos, etc.), the more difficult is to generate realistic simulated data by means of parametric methods.

In this framework, generative AI methods opens a completely new range of possibilities, as they can augment the capacity to replicate and generate data patterns that resemble real ones regardless of their complexity.

This study reviews generative AI architectures for synthetic data generation and their current state of the art in manufacturing applications. Special attention is devoted to a sub-class of methods known as generative adversarial networks (GANs), because of their flexibility in dealing with a variety of industrially relevant problems. We review and critically discuss the various open issues and limitations that deserved additional research and industrial developments.

2 Generative AI Architectures and Methods

Generative AI methods for synthetic data generation include different architectures and training paradigms, namely: variational autoencoders (VAEs), normalizing flows, diffusion models, autoregressive models and GANs. In terms of data format, the range of applications and methods is broad, ranging from high-dimensional multivariate data to time series [5, 6], from image data to videos [7]. All of them are relevant for manufacturing applications, as in-line signals acquired during the process can be in various formats. In this study we focus mainly on image data, as the number of industrial applications involving machine vision techniques have increased substantially in the recent years in high precision manufacturing and additive manufacturing [8, 9].

VAEs use an encoder-decoder architecture to learn the latent representations of input data [10]. The encoder maps the input data to a latent space, while the decoder generates new data samples by drawing from the probability distribution learned in the latent space and back-transforming to the original data domain. This approach is suitable to generate synthetic data (e.g., images) whose latent structure is difficult to model. VAEs can draw new samples from an image dataset, but users have limited control on the specific features they can reproduce in synthetic data, e.g., anomalous patterns of interest. *Normalizing flows* consist of series of simple and invertible functions that are sequentially applied to the input data to transform a complex image pattern into a simple probability distribution [11]. Random samples are then drawn from the simple distribution to obtain new images via backward application of serial functions. They suffer from similar limitations of VAEs in terms of flexibility. *Diffusion models* rely on a different way to implement forward and backward transformations [12]. The forward (diffusion) process gradually adds Gaussian noise to input images, whereas the reverse (generative) process starts from pure noise to generate new image samples. This approach has recently gained popularity due to their combination with large language models (LLMs) to generate images based on text prompts. *Autoregressive models* can treat images as sequences of pixels and predict pixel intensities across the vertical and horizontal axes [13]. This is achieved by modelling the conditional probability distribution of pixel intensities based on the occurrence of previous ones. They have the potential to reproduce specific patterns and transfer image features from real sample to synthetic ones, but the way to enhance the output quality is still an area of research.

GANs represent another different category of generative AI models, which has attracted increasing interest in industrial applications [14]. They consist of two neural networks known as the "generator" and the "discriminator", respectively, which are trained simultaneously in a competitive process (Fig. 1).

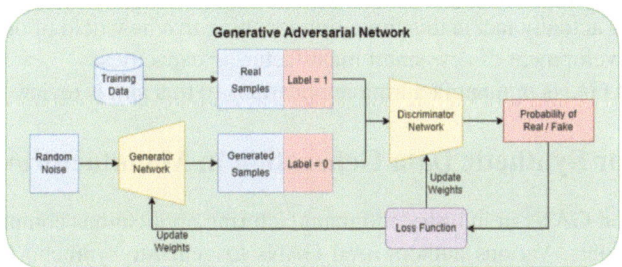

Fig. 1. General scheme of a GAN architecture [14]

The *generator* creates synthetic images starting from random noise generating patterns that the discriminator fails in distinguishing from real data, thus maximizing the discriminator error rate. The *discriminator* receives as inputs both synthetic and real data aiming to minimize its classification error. They are trained together in a minimax game. A single loss function, e.g., the Jensen-Shannon divergence [14], is used to optimize both.

Basic GANs, as well as other generative methods, do not provide any direct control on the ability of the network to capture salient defect features. Some GAN architectures have been proposed to specifically enhance the control on the type of generated images, opening a range of new possible uses of these methods. Three examples are:

- *Conditional GANs* (cGANs): they condition the generation process on additional information, e.g., class labels which aid the generation of synthetic data in separate classes, preserving the between class dissimilarities.
- *Style transfer GANs* (styleGANs): they perform image-to-image translation, i.e., they transfer patterns learned in one type of image, e.g., the one including defects, to the other type, e.g., the defect-free image.
- *Cycle consistency GANs* (cycleGANs): they perform a similar pattern transfer, but using a loss function that preserves key characteristics of the original image in the translated one.

The benefits of generative AI methods, and GANs in particular, with respect to non-generative AI approaches for synthetic data generation have been discussed by various authors [15]. As an example, [16, 17] developed a GAN-based approach to generate synthetic microstructure data, showing that compared to the commonly used microstructure generation algorithms, the GAN results are structurally and electrochemically more realistic. Similar conclusions were drawn in other applications, e.g., in generating melt pool and weld pool images under unexplored experimental conditions [18].

Generally speaking, data simulation has been widely used in the literature to train and/or test machine learning methods. However, in manufacturing applications, simulating realistic data patterns (e.g., product defects and process anomalies) in the presence of highly complex process dynamics represents a challenging task. Simulated patterns are commonly hardly representative of the complex spatial and/or temporal features of real phenomena. As a result, only macroscopic and simplified anomalies can be simulated in most cases. Synthetic data generation via GANs and other generative AI methods has the potential to actually tackle this limitation, opening to a new field of opportunities to support the development of new smart manufacturing capacities.

The use of GANs in manufacturing applications to this aim is reviewed in Sect. 3.

3   GANs for Synthetic Data Generation in Manufacturing

The research on GANs in industry and manufacturing applications comprises a variety of seminal studies. Various authors used GANs to generate synthetic microstructure data [16, 17]. Microstructure data are time-consuming and expensive to collect, which results in limited and unbalanced datasets. This represents a limitation to properly train machine learning classifiers. Simulation software tools have been developed to cope with this issue, but some authors showed that microstructure images generated via GANs are more "structural and electrochemically realistic" than competitor methods [17].

A fully convolutional GAN was proposed in [16] where they used two input parameters, namely a seed to add stochastic variations and morphological parameter that controls the morphological characteristics of the images (grain size, orientation, etc.). Thanks to these parameters they could achieve an accurate control in the generation of desired patterns (Fig. 2).

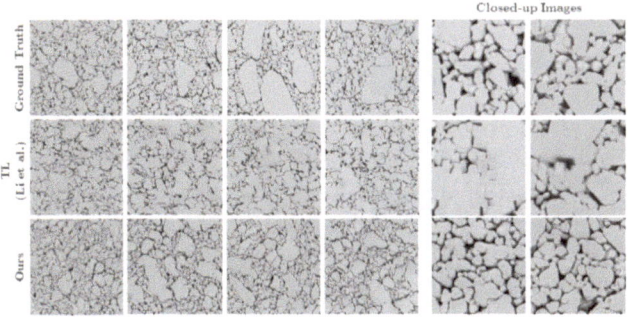

Fig. 2. Example of real (ground truth) and synthetic microstructure images from [16]

Another industrial application regards the generation of synthetic defect images. Singh et al. [19] compared styleGANs and cycleGANs to this aim in a semiconductor manufacturing application. The major difference between them is that styleGANs are constrained to transfer patterns among paired images, while cycleGANs do not have this limitation. Since images of defects are commonly more scarcely available than images in normal conditions, cycleGANs are much more flexible for industrial adoption. Singh et al. [19] also showed that cycleGANs outperformed styleGANs in terms of synthetic image quality.

Zhang et al. [20] developed a variant of cGANs for paired images to generate defects on a clean, defect-free fabric image (Fig. 3). A pre-process steps was included to select the defective region, and a second stage was added to fine-tune the fabric defect image. The authors trained different object detection algorithms using as input both real and synthetic data. They showed that augmenting the training with synthetic images improved the final object detection performances. Similar results were confirmed by other authors [21–25].

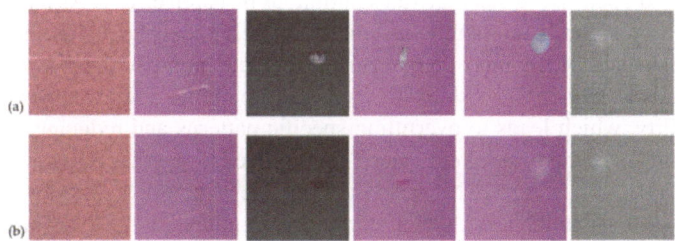

Fig. 3. Example of synthetic defect images after the first (a) and second (b) GAN stage in [13]

Being able to control defect location and size or severity is a key aspect to generate valuate synthetic defect data. A method aimed to enhance this capability have been proposed so far. Niu et al. [26], where synthetic defect locations were controlled by using defect masks as inputs to the GAN. However, the flexibility in replicating defects with desired properties still represents an open issue, as also shown in other variants of GAN architectures applied to machine vision inspection applications [27, 28].

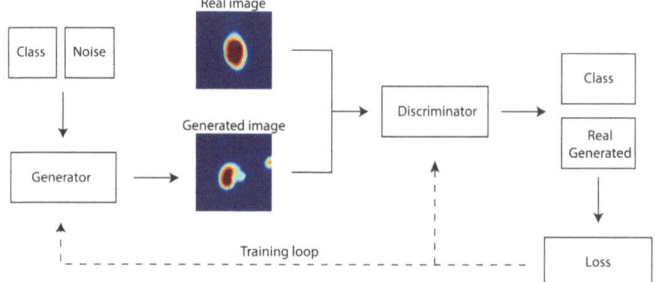

Fig. 4. cGAN approach for synthetic generation of melt pool images [18]

In some cases, authors leveraged on the synthetic data generation capability to develop GAN-based defect detection or classification methods [18, 29]. In [18] a conditional GAN was proposed to generate melt pool images in metal AM and the same network was used in-line to classify in-situ gathered melt pool image streams (Fig. 4). The cGAN architecture was fed with labelled images (namely video frames treated as independent images, thus ignoring the temporal dependence), where labels indicated the process parameters associated to every melt pool image. The discriminator was trained not only to distinguish between real and synthetic data, but also to estimate the label of every input image, either real or synthetic. Petrik et al. [18] also showed that the conditional training enabled the extrapolation of melt pool patterns under experimental conditions that were not included in the training.

4   Open Issues and Perspectives

Despite the several benefits, generative AI methods are still characterized by various limitations and open issues that motivate continuous research efforts. As far as GANs are concerned, they suffer from training instability, namely the difficulty in finding a stable equilibrium during the minimax game between the generator and the discriminator, and "mode collapse", namely the risk for the generator to ignore a relevant portion of input data variability, which leads to overfitting specific patterns and reducing the synthetic data variability with respect to real data. The improved performances enabled by some types of GANs, e.g., cycleGANs, also came at the expense of a more complicated architecture (more parameters), which commonly implies the need for larger training datasets.

One critical aspect regards the lack of methods to 1) quantify the GAN's progress during the training and 2) validate the quality of its generated outputs. GAN's architectural complexity makes the training unstable, which makes convergence difficult or even impossible to achieve. Lack of convergence and mode collapse have a detrimental effect on the final output. It also makes difficult to tune the network through hyperparameter optimization. Despite various attempts to tackle this limitation, it still represents an open issue [30].

Regarding the validation of generated outcomes, some metrics have been proposed so far. They include: the Fréchet inception distance (FID), the inception score, the average log-likelihood, the multi-scale structural similarity (MS-SSIM) and the perceptual image patch similarity (LPIPS). Some authors used such metrics to compare alternative methods [27], but they suffer from several drawbacks. In some cases, they are based on heuristics that does not necessary capture salient features of interest, in other cases they rely on external neural networks and need very large dataset to provide reliable results [31]. Such limitations still make visual inspection the most effective validation approach, but this represents a barrier for the wide adoption of these techniques.

Another important issue regards the extrapolation limitations. GANs, as most generative AI methods, replicate data samples that resembles input data patterns, without direct control on the type, location and properties of the features that shall be included or excluded from generate data. This means that such techniques are limited in the way in which they can extrapolate specific patterns and generate variants that vary in size, intensity, location, etc. To some extents, this issue has been recently addressed with conditional GANs, e.g., by embedding morphological parameters, and combining them with pre-processing operations or fine-tuning stages. However, this approach provides some control on the global properties of the image, not on specific regions of interest. This represents a field where additional research is needed.

Other promising directions involve novel GAN architectures explored in few seminal studies to extend their capabilities moving from image data to more complex domains. Two examples include GANs for synthetic video generation, where both spatial and temporal patterns are captured and reproduced in a realistic way, and 3D GANs suitable to deal with three dimensional objects.

Aforementioned challenges still represent barriers for the industrial adoption of GANs in real manufacturing environments. Although various real case studies have been presented and successfully addressed in the scientific literature, the technology readiness level (TRL) in the specific field of quality monitoring and control is still low, between TRL 3 and 5. Most of the literature focused on open datasets, whereas only few authors tested the proposed methods in relevant environments using real data. The lack of a solid validation framework still imposes subjective evaluations that in most cases are not compliant with stringent quality requirements and implementation constraints. Moreover, the difficulty in controlling the convergence of trained models and the lack of interpretability underpin the general caution exercised by early adopters towards these methods. This motivates continuous research efforts not only understand the potentials of the methods but also to provide industry with trustworthy solutions to highly complex problems.

Moreover, synthetic data generation requires high processing power and large data volumes. Cloud computing infrastructures are commonly used for network training, but this also opens cybersecurity issues that shall be faced for effective industrial adoption. Not all potential adopters already have internal infrastructures compatible with the introduction and integration of such novel tools. This pushes the need to bridge technological and cultural gaps to leverage the benefits provided by AI and generative AI in the digital transition. Federated learning combined with generative AI may be one way to unlock the potential of new advanced models while meeting security requirements. Seminal solutions in this direction have been already explored [32].

Finally, the potential GANs is not limited to data generation. They can also be used for anomaly detection and classification. Generative models can be trained to learn the normal patterns of industrial processes and generate "in-control" representations, such that any deviation from these learned patterns can be signaled as alarms, aiding in-line anomaly detection. Moreover, they can predict the outcomes of different process parameters, allowing for adjustments that may enhance the quality and performance of the product. In this fields, GANs can be combined with statistical methods to tackle complex problems opening to a range of new application scenarios.

5 Conclusion

Most cited studies were focused on testing GANs on open datasets, while the number of studies addressing real use cases in manufacturing applications is still quite limited. Such seminal studies demonstrated the high potential of the methods, but they also highlighted a broad range of open issues and challenges that deserve additional research efforts. It is also worth pointing out that the use of generative AI methods different from GANs is much less explored, despite the notable performances and capacities they exhibited in other applications. Generative AI has the potential to trigger new opportunities for the development of advanced solutions in smart and zero-defect manufacturing. This represents a low maturity field compared to the adoption of other machine learning methods, several challenges shall be faced, and new solutions are needed, but the research is progressing at a quite fast pace.

Acknowledgements. This study was partially funded by MICS (Made in Italy – Circular and Sustainable) Extended Partnership received funding from the European Union Next-GenerationEU (PIANO NAZIONALE DI RIPRESA E RESILIENZA (PNRR) – MISSIONE 4 COMPONENTE 2, INVESTIMENTO 1.3 – D.D. 1551.11-10-2022, PE00000004).

References

1. Global Artificial Intelligence in Manufacturing Market Size, Share & Industry Trends Analysis Report By Offering (Software, Hardware, and Services), By Application, By Technology, By Industry, By Regional Outlook and Forecast, 2022 – 2028 (2022). https://www.kbvresearch.com/artificial-intelligence-in-manufacturing-market/
2. Decardi-Nelson, B., Alshehri, A.S., Ajagekar, A., You, F.: Generative AI and process systems engineering: the next frontier. Comput. Chem. Eng. **187**, 108723 (2024)
3. Jo, A.: The promise and peril of generative AI. Nature **614**(1), 214–216 (2023)
4. Khan, S.S., Madden, M.G.: One-class classification: taxonomy of study and review of techniques. Knowl. Eng. Rev. **29**(3), 345–374 (2014)
5. Zhang, C., Kuppannagari, S.R., Kannan, R., Prasanna, V.K.: Generative adversarial network for synthetic time series data generation in smart grids. In: 2018 IEEE International Conference on Communications, Control, and Computing Technologies for Smart Grids (SmartGridComm), pp. 1–6. IEEE (2018)
6. Brophy, E., Wang, Z., She, Q., Ward, T.: Generative adversarial networks in time series: a systematic literature review. ACM Comput. Surv. **55**(10), 1–31 (2023)

7. Aldausari, N., Sowmya, A., Marcus, N., Mohammadi, G.: Video generative adversarial networks: a review. ACM Comput. Surveys (CSUR) **55**(2), 1–25 (2022)
8. Grasso, M., Remani, A., Dickins, A., Colosimo, B.M., Leach, R.K.: In-situ measurement and monitoring methods for metal powder bed fusion: an updated review. Meas. Sci. Technol. **32**(11), 112001 (2021)
9. Gao, W., et al.: On-machine and in-process surface metrology for precision manufacturing. CIRP Ann. **68**(2), 843–866 (2019)
10. Doersch, C.: Tutorial on variational autoencoders (2016). arXiv:1606.05908
11. Kobyzev, I., Prince, S.J., Brubaker, M.A.: Normalizing flows: an introduction and review of current methods. IEEE Trans. Pattern Anal. Mach. Intell. **43**(11), 3964–3979 (2020)
12. Zhang, C., Zhang, C., Zhang, M., Kweon, I.S.: Text-to-image diffusion model in generative AI: a survey (2023). arXiv:2303.07909
13. Bond-Taylor, S., Leach, A., Long, Y., Willcocks, C.G.: Deep generative modelling: a comparative review of vaes, gans, normalizing flows, energy-based and autoregressive models. IEEE Trans. Pattern Anal. Mach. Intell. **44**(11), 7327–7347 (2021)
14. Goodfellow, I., et al.: Generative adversarial networks. Commun. ACM **63**(11), 139–144 (2020)
15. Eigenschink, P., Reutterer, T., Vamosi, S., Vamosi, R., Sun, C., Kalcher, K.: Deep generative models for synthetic data: a survey. IEEE Access **11**, 47304–47320 (2023)
16. Chun, S., Roy, S., Nguyen, Y.T., Choi, J.B., Udaykumar, H.S., Baek, S.S.: Deep learning for synthetic microstructure generation in a materials-by-design framework for heterogeneous energetic materials. Sci. Rep. **10**(1), 13307 (2020)
17. Hsu, T., et al.: Microstructure generation via generative adversarial network for heterogeneous, topologically complex 3d materials. JOM **73**, 90–102 (2021)
18. Petrik, J., Kavas, B., Bambach, M.: MeltPoolGAN: auxiliary classifier generative adversarial network for melt pool classification and generation of laser power, scan speed and scan direction in Laser Powder Bed Fusion. Addit. Manuf. **78**, 103868 (2023)
19. Singh, R., Garg, R., Patel, N.S., Braun, M.W. Generative adversarial networks for synthetic defect generation in assembly and test manufacturing. In: 2020 31st Annual SEMI Advanced Semiconductor Manufacturing Conference (ASMC), pp. 1–5. IEEE (2020)
20. Zhang, Y., et al.: A novel defect generation model based on two-stage GAN. e-Polymers **22**(1), 793–802 (2022)
21. Zhu, K., Chen, W., Hou, Z., Wang, Q., Chen, H.: Modified fusing-and-filling generative adversarial network-based few-shot image generation for GMAW defect detection using multi-sensor monitoring system. Int. J. Adv. Manuf. Technol. **128**(5–6), 2753–2762 (2023)
22. Niu, S., Li, B., Wang, X., Lin, H.: Defect image sample generation with GAN for improving defect recognition. IEEE Trans. Autom. Sci. Eng. **17**(3), 1611–1622 (2020)
23. Ji, Y., Lee, J.H.: Using GAN to improve CNN performance of wafer map defect type classification: yield enhancement. In: 2020 31st Annual SEMI Advanced Semiconductor Manufacturing Conference (ASMC), pp. 1–6. IEEE (2020)
24. He, X., Luo, Z., Li, Q., Chen, H., Li, F.: DG-GAN: a high quality defect image generation method for defect detection. Sensors **23**(13), 5922 (2023)
25. Gong, Y., Liu, M., Wang, X., Liu, C., Hu, J.: Few-shot defect detection using feature enhancement and image generation for manufacturing quality inspection. Appl. Intell. **54**(1), 375–397 (2024)
26. Niu, S., Li, B., Wang, X., Peng, Y.: Region-and strength-controllable GAN for defect generation and segmentation in industrial images. IEEE Trans. Ind. Inf. **18**(7), 4531–4541 (2021)
27. Ran, G., Yao, X., Wang, K., Ye, J., Ou, S.: Sketch-guided spatial adaptive normalization and high-level feature constraints based GAN image synthesis for steel strip defect detection data augmentation. Meas. Sci. Technol. **35**(4), 045408 (2024)

28. Su, B., Zhou, Z., Chen, H., Cao, X.: SIGAN: a novel image generation method for solar cell defect segmentation and augmentation (2021). arXiv:2104.04953
29. Zhang, L., Dai, Y., Fan, F., He, C.: Anomaly detection of GAN industrial image based on attention feature fusion. Sensors **23**(1), 355 (2022)
30. Saxena, D., Cao, J.: Generative adversarial networks (GANs) challenges, solutions, and future directions. ACM Comput. Surveys **54**(3), 1–42 (2021)
31. Elgendy, M.: Deep Learning for Vision Systems. Simon and Schuster (2020)
32. Cao, X., Sun, G., Yu, H., Guizani, M.: PerFED-GAN: personalized federated learning via generative adversarial networks. IEEE Internet Things J. **10**(5), 3749–3762 (2022)

Laser Metal Deposition (LMD) Process Monitoring: From 3D Visualization of Sensor Data Towards Anomaly Detection

Mikel Ayuso[1(✉)], Ander Muniategui[1], Aitor Aguirre-Ortuzar[2], and Enaitz Ezpeleta[2]

[1] LORTEK, Basque Research and Techonology Alliance (BRTA),
Arranomendia kalea 4A, 20240 Ordizia, Spain
mayuso@lortek.es

[2] Electronics and Computing Department, Mondragon Unibertsitatea,
Mondragon, Gipuzkoa, Spain

Abstract. Metal Additive Manufacturing (AM) allows producing geometrically complex metal components, unlocking new design possibilities and making it suitable to sectors such as healthcare, automotive and aerospace. AM processes are complex and require the use of many sensors to extract relevant process information for its monitoring and control. In the last years, many studies have applied advanced Deep Learning methods to extract knowledge from AM processes. However, these developments are specific to a particular setup, problem or defectology. Furthermore, they lack frameworks and pipelines to guide throughout their development, and do not include AI-related tools for data labelling, visualization, and AI model development and deployment. With the aim of simplifying the development and deployment of AM process monitoring systems, a dashboard-based framework that makes use of AI for anomaly detection and for feature extraction is presented in this study. The framework helps with development and deployment of monitoring systems by easing the incorporation of new sensors and the extraction of new features from captured data by end users. In this study, a Laser Metal Deposition (LMD) process is considered as the use case to show the usefulness of the developed framework.

Keywords: laser metal deposition · process optimization · 3D visualization · anomaly detection

1   Introduction

Metal Additive Manufacturing (AM), also known as metal 3D printing, has transformed the manufacturing sector with its remarkable flexibility, accuracy and its capacity to manufacture with challenging materials. Embraced across diverse industries including aerospace, automotive, and healthcare, AM fuels

K. Alexopoulos et al. (Eds.): ESAIM 2024, LNME, pp. 31–39, 2025.
https://doi.org/10.1007/978-3-031-86489-6_4

innovation crafting or fixing three-dimensional metal components by layering materials and unlocking new design possibilities [1]. The lack of closed loop control and highly dynamic nature of AM processes make high quality consistency and process repeatability challenging [2]. Numerous parameters and physical effects such as input energy, gas flow, melt-pool characteristics, welding seam geometry, heat transfer between layers, and others must be managed to prevent defects like Lack of Fusion (LoF), pores, deformations, and cracks [3]. AM process monitoring, control and post-process quality control requires complex multi sensor setups. For instance, CCD cameras, welding cameras, thermal cameras, profilometers, acoustic emission among others have been used. Implementation of these setups is not straightforward and implies sensor positioning, calibration, data capture and storage, data synchronization, sensor data analysis, multi-sensor data analysis and final integration of developments, including trained AI models, among others. Furthermore, optimization of parameters via trial-and-error mechanistic modelling or via machine learning tools are time-consuming, and it does not warranty the quality consistency [4].

Different AI-based approaches have been used to extract process knowledge, detect defects, monitor or control AM processes. Among the different AI-based approaches, anomaly detection and deep learning-based methods are the most extended ones [4]. AI-based solutions tend to be use case and setup specific and lack of an integral tool that helps the user throughout the entire pipeline from data capture to model deployment and final process monitoring and control. The developed framework contains a user-friendly UI-based interface designed to help end users to quickly identify deviations and adjust the system. It integrates a set of tools that together with a reference pipeline serves to adjust the setup and the LMD process. Finally, data is visualized in 3D with to identify the position and type of deviations and defects. Its usefulness is shown within LMD trials in simple geometries. Powder flow and humping defects are detected via anomaly detection.

2   Background

Laser Metal Deposition (LMD), also known as Laser Cladding or Direct Metal Deposition, is a directed energy deposition class of AM technology that consists of the deposition of metal powder or wire onto a substrate using a high-energy laser beam [5]. LMD is a versatile and progressively favoured AM method [1]: material deposition can be precisely regulated, it allows the printing of complex geometries with minimal material loss, and it can be used to mend valuable components. Conversely, it remains a relatively novel AM technology, with some aspects of its underlying physics still awaiting full comprehension [4]. Gas porosity (voids or gas pockets within the deposited material), lack of fusion (inadequate bonding between the deposited material and the substrate or layers), surface irregularities and humping (undulations along the printed layer's surface) are among the prevalent imperfections occurring in LMD. In the particular case of humps, their appearance can be due to high scan speeds, or the relation between input energy, amount of powder deposited and laser interaction time. These deviations can get exponentially worse in subsequent layers, leading to defects such as porosity or lack of fusion [6].

2.1 Laser Metal Deposition Process Monitoring and Control

Process monitoring in industrial settings is crucial for ensuring product quality, optimizing efficiency, and maintaining safety standards. Among the different monitoring methods, the most common technique is in-situ sensor monitoring where sensors are embedded within the manufacturing equipment to collect real-time data, i.e. temperature, pressure, material deposition rates, and acoustic emissions [7]. This approach provides immediate feedback to operators, allowing them to make timely adjustments to process parameters and prevent defects. Welding and thermal cameras [8], acoustic sensors and profilometry are the most widespread sensors used in LMD process monitoring [4].

Thermal cameras capture infrared radiation emitted by the heated material and surrounding components, provide visual representation of temperature distributions across the piece and enables the identification of hotpots, monitor heat dissipation, and detect lack of fusion or porosity. However, the spatial resolution of thermal cameras may limit their ability to detect subtle variations in temperature. Additionally, welding cameras are used to extract additional information, such as, geometry of the newly printed area, melt-pool, layer thickness and humping formation. Due to their high spatial resolution, welding cameras are extensively used for LMD process monitoring [4].

Despite the advancements in process monitoring and control techniques [9,10], several challenges persist in industrial applications, including the integration of multiple monitoring methods into a unified monitoring system, the development of robust algorithms for real-time anomaly detection, and the implementation of remote monitoring capabilities for decentralized manufacturing facilities [4].

2.2 AI Tools: Deep Learning Models, Data Visualization and Labelling

Data visualization and UIs play a crucial role in industrial processes, as they are the link between the industrial components and the workers, particularly in cutting-edge fields like AM and LMD [4]. In recent years, a big leap is being developed in regard to UIs and user experience [11]. These innovative techniques rely on precise data analysis and visualization to ensure quality control, process optimization, and performance monitoring. Several methods are employed to visualize data effectively. Through the use of UIs and dashboards, operators can access comprehensive displays that show images of the deposition process captured by high-resolution cameras. These images provide real-time feedback on factors such as powder distribution, surface quality, and melt-pool dynamics [4]. In addition to image data, sensor process variables such as temperature, pressure, and laser power are continuously monitored and visualized in the UI.

Not all cameras include their own image analysis software and hence, development and deployment of image analysis models for feature extraction it is an extended procedure. DL-based models require large amounts of labelled images, for manual image annotation, different tools like Labelbox [12] or COCO Annotator [13] can be used to speed the process. Semi-automated labelling, where certain parts of the process are automated [14], reducing human intervention to annotation revision and correction, eases this process even further. Completely automated labelling is also possible, usually this approach is used with some kind of clustering algorithm [15]. However, these tools are not, in general, integrated within a unique framework that integrates AM process monitoring, with visualization, data filtering and labelling, and DL-model training and deployment.

3   Developments

Setup definition for monitoring and close loop control continues to be an open task for metal AM processes. As new sensors appear and existing sensors evolve by their miniaturization or to stand harsh environments, setup definition, trial and adjustments will need to be faster, more flexible and simpler. The aim of this work is to ease and speed up setup implementation, to accomplish this, a UI-based framework with the following main features has been developed: 1) *flexibility* to be used with different kind of sensor and captured data types, 2) *user-friendly*, helpful for end-users at trial stages of setup definition, and 3) *easy to use*, with simple steps that can be executed in an interface that follows a simple pipeline developed for sensor configuration and setup implementation.

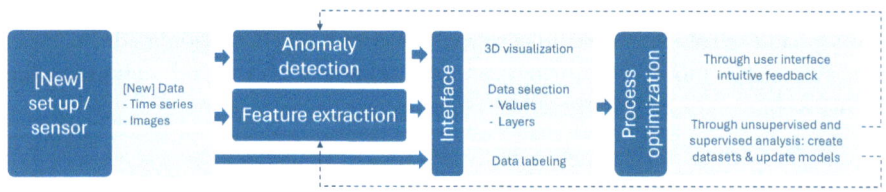

Fig. 1. Pipeline for sensor data configuration for monitoring.

In this study, a laboratory level implementation for LMD setup configuration is considered. Synchronization of captured data is out of scope of this article. The focus is placed on single sensor configuration and integration, data capture, visualization and anomaly detection. This study aims at laying the foundations of a future, more extensive tool for multisensor data analysis for process control. In the following subsections, the reference pipeline and an implementation of the use of the framework for a welding camera for an LMD process are described.

3.1   Pipeline: Sensor Data Monitoring

A three-step iterative loop pipeline to extract relevant data from each sensor has been defined (Fig. 1). The first step is to extract process information from captured data by using unsupervised or supervised ML or DL models. While some of the sensors do incorporate their own software for process parameter extraction, many do not, in those cases, the use of unsupervised models is a cost-effective alternative to filtering captured data and generating proper datasets to fed supervised models. Among unsupervised ones, the most extended ones are anomaly detection models. Once the model is trained, any data outside the normal behaviour will be scored as abnormal. Anomaly detection on time series and images is carried out following similar steps. For instance, in the case of images, the user can specify the region of interest (ROI), this is useful in cases where the images contain relevant information about different parameters of the process. For example, images where the nozzle and part of the printed part is shown. In these cases, several ROI could be defined. In the second step, anomaly detection results are mapped into a 3D representation of the printed part. This allows filtering by layer and/or anomaly scores, with filtered values displayed in a new window for manual labelling to verify detection accuracy. Outputs from supervised models can also be visualized and annotated, this labelled data is used to adjust anomaly detection models, as well as supervised models for classification or feature extraction using object detection or segmentation models. The last step of the pipeline consists of adding / updating the DL-based models within the real time process monitoring. This step closes the loop of the pipeline and serves to adjust process parameters from observation through end-users. In the next iterations of the pipeline, 3D visualization is used for real time data monitoring of raw values, extracted features and detected anomalies.

3.2   User-Interface Based Framework

The developed framework unifies DL-based process analysis with visualization technology (Fig. 2.c and 2.e). It allows data to be visualized in an easy and intuitive format, and enables easy data labelling for model training. 3D environments and interfaces are key in AM and LMD processes, specially during the part design process, by continuing development of 3D visualization and UI tools, the job of design engineers can be improved and optimized [16], this can be also applied to the manufacturing and post-processing part of the process.

A new custom software using direct calls to OpenGL is capable of smoothly showing up to 5 million data points in a 3D virtual space. The user is also able to navigate the 3D space freely, rotate the virtual representation of the workpiece and filter the visible data by layer, either showing all the workpiece, a single layer of the manufacturing process or a combination of various layers. By visualizing the manufacturing process in a 3D space, engineers and technicians can identify potential flaws or areas of improvement in real-time. This approach enables adjustments to be made quickly, reducing the likelihood of errors and minimizing material wastage.

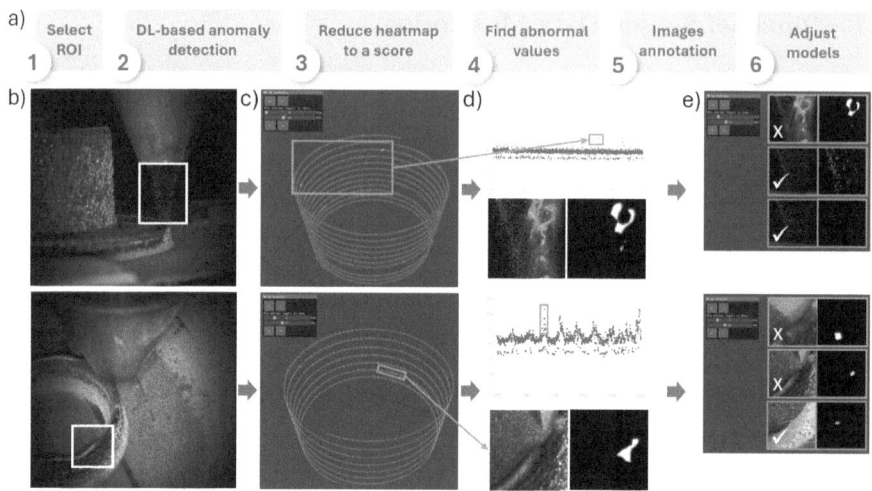

Fig. 2. Developed framework and pipeline for a welding camera for LMD process.

3.3 LMD Use Case: Welding Camera and Anomaly Detection

The integration of a welding camera within an LMD monitoring setup printing a simple spiral geometry is described here. The setup consists of a Cavitar C300 welding camera, a Wenglor MLSL143 Profilometer, and a NIT Clamir thermographic camera. The welding camera, placed off axis, enables real-time, high-resolution observation of the laser-metal interaction during deposition. The Wenglor MLSL143 Profilometer, placed on axis with an offset just after the nozzle, adds precise topographical data measurement capabilities. Finally, the NIT Clamir system, on axis, gives insight about temperature distribution, thermal behaviour and melt-pool dynamics. Two positions were tested for the welding camera: One for analysing the powder flow (Fig. 2.b top), and the other one for determining the geometry of the newly welded area (Fig. 2.b down). In order to monitor the powder flow and geometry of the newly welded area, the pipeline described in the previous section is applied as indicated in Fig. 2. The first step consists of determining a ROI for each of the images (red boxes on Fig. 2.b). The ROI is extracted from all captured images and those related with normal process behaviour are used to train DL-based anomaly detection. PaDiM anomaly detection [17] is used in this second step, which outputs a heatmap where regions with higher score are indicative of anomaly regions. This output is visually inspected to update the training dataset and to generate a new dataset that could be used for alternative DL-based models. In the third step, the output heatmap images are reduced into a unique score to simplify filtering and selection for labelling. Different image scoring options have been tried: based on percentiles (50, 75 and 95) and based on delentropy [18], which is an extended version of the Shannon information for 2D images, a measure of the complexity of image data [19]. The fourth step consists of visualizing these scores in 3D representing the geometry of the printed part (Fig. 2.c). The equivalent 2D visualization of data together with original and heatmap images of two detected anomalies are shown in Fig. 2.d. Scores are then filtered by layer and/or value, and their original and heatmap images are loaded in a new window for manual labelling (Fig. 2.e). This is carried out in the fifth step just by click the images the algorithm has wrongly identified as anomalies. Finally, in the sixth step, manually curated data is used to adjust PaDiM or alternative DL-based models that are incorporated in posterior iterations.

4 Conclusions and Future Steps

To ease setup definition and process monitoring, a user-friendly framework has been developed, tested in a laboratory setup for LMD process monitoring for anomaly detection, and presented in this document. The tool has two major parts: an interface and a reference pipeline. The interface consist of a principal window for the visualization of data in a 3D plot with the geometry of the printed part, and a secondary window for the labelling of images to identify predicted False Positives (FP). Easy and intuitive data visualisation enables fast reaction times from technicians and the adjustment of DL-based models. Finally, the iterative pipeline guides the development of aforementioned DL-based models. Currently, the annotation is restricted to a binary classification to identify FP from model output that serve to generate datasets to train anomaly detection as well as classification models. In the future, the aim is to extend the annotation by making use of alternative available annotations tools (COCO-Annotator and Segment Anything models for example) so that other types of models, such as object detection or segmentation, can be trained and used to determine the values of other relevant process parameters. Furthermore, additional tools for time series analysis will be incorporated progressively. This framework has been developed to be user-friendly, help in fast definition and implementation of monitoring setups, and aimed towards a process control.

References

1. Bruzzo, F., et al.: Sustainable laser metal deposition of aluminum alloys for the automotive industry. J. Laser Appl. **34** (2022). https://doi.org/10.2351/7.0000741
2. Ouidadi, H., Guo, S., Zamiela, C., Bian, L.: Real-time defect detection using online learning for laser metal deposition. J. Manuf. Process. **99**, 898–910 (2023)
3. Herzog, T., Brandt, M., Trinchi, A., Sola, A., Molotnikov, A.: Process monitoring and machine learning for defect detection in laser-based metal additive manufacturing. J. Intell. Manuf. **35** (2024). https://doi.org/10.1007/s10845-023-02119-y
4. Chen, L., et al.: In-situ process monitoring and adaptive quality enhancement in laser additive manufacturing: a critical review. J. Manuf. Syst. **74**, 527–574 (2024)
5. Li, F., Gao, Z., Li, L., Chen, Y.: Microstructural study of MMC layers produced by combining wire and coaxial WC powder feeding in laser direct metal deposition. Opt. Laser Technol. **77**, 134–143 (2016). https://doi.org/10.1016/j.optlastec.2015.09.018

6. Caprio, L., Demir, A.G., Previtali, B.: Observing molten pool surface oscillations during keyhole processing in laser powder bed fusion as a novel method to estimate the penetration depth. Add. Manuf. **36** (2020). https://doi.org/10.1016/j.addma.2020.101470
7. Prieto, C., et al.: 11 th CIRP conference on photonic technologies (2020)
8. D'accardi, E., et al.: Monitoring the laser metal deposition (LMD) process by means of thermal methods. QIRT Council (2022). https://doi.org/10.21611/qirt.2022.2014
9. Ngoc Vu Nguyen, Allen Jun Wee Hum, T.D., Tran, T.: Semi-supervised machine learning of optical in-situ monitoring data for anomaly detection in laser powder bed fusion. Virtual Phys. Prototyping **18**(1), e2129396 (2023). https://doi.org/10.1080/17452759.2022.2129396
10. García-Díaz, A., et al.: OpenLMD, an open source middleware and toolkit for laser-based additive manufacturing of large metal parts. Robot. Comput.-Integr. Manuf. **53**, 153–161 (2018)
11. Nee, A.Y., Ong, S.K., Chryssolouris, G., Mourtzis, D.: Augmented reality applications in design and manufacturing. CIRP Ann. Manuf. Technol. **61**, 657–679 (2012). https://doi.org/10.1016/j.cirp.2012.05.010
12. Labelbox: labelbox. https://labelbox.com (2024)
13. Brooks, J.: COCO Annotator. https://github.com/jsbroks/coco-annotator/ (2019)
14. Larumbe-Bergera, A., Porta, S., Cabeza, R., Villanueva, A.: SeTA: semiautomatic tool for annotation of eye tracking images. Assoc. Comput. Mach. (2019). https://doi.org/10.1145/3314111.3319830
15. Sager, C., Janiesch, C., Zschech, P.: A survey of image labelling for computer vision applications. J. Bus. Anal. **4**, 91–110 (2021). https://doi.org/10.1080/2573234X.2021.1908861
16. Rodriguez-Conde, I., Campos, C.: Towards customer-centric additive manufacturing: making human-centered 3D design tools through a handheld-based multi-touch user interface. Sensors **20**, 1–28 (2020). https://doi.org/10.3390/s20154255
17. Defard, T., Setkov, A., Loesch, A., Audigier, R.: PaDiM: a patch distribution modeling framework for anomaly detection and localization (2020). http://arxiv.org/abs/2011.08785
18. Larkin, K.G.: Reflections on Shannon information: in search of a natural information-entropy for images structure of paper (2016)
19. Wu, W., Daneker, M., Turner, K., Jolley, M., Lu, L.: Identifying heterogeneous micromechanical properties of biological tissues via physics-informed neural networks (2024)

An Explainable Active Learning Approach for Enhanced Defect Detection in Manufacturing

Nikolaos Nikolakis, Paolo Catti, and Kosmas Alexopoulos[✉]

Laboratory for Manufacturing Systems and Automation, Mechanical Engineering and Aeronautics Department, University of Patras, 26504 Patras, Greece
alexokos@lms.mech.upatras.gr

Abstract. Artificial Intelligence (AI) can significantly support manufacturing companies in their pursuit of operational excellence, by maintaining efficiency while minimizing defects. However, the complexity of AI solutions often creates a barrier to their practical application. Transparency and user-friendliness should be prioritized to ensure that the insights generated by AI can be effectively applied in real-time decision-making.

To bridge this gap and foster a collaborative environment where AI and human expertise collectively drive operational excellence, this paper suggests an AI approach that targets identifying defects in production while providing understandable insights. A semi-supervised convolutional neural network (CNNs) with attention mechanisms and Layer-wise Relevance Propagation (LRP) for explainable active learning is discussed. Predictions but also feedback from human experts are used to dynamically adjust the learning focus, ensuring a continuous improvement cycle in defect detection capabilities. The proposed approach has been tested in a use case related to the manufacturing of batteries. Preliminary results demonstrate substantial improvements in prediction accuracy and operational efficiency, offering a scalable solution for industrial applications aiming at zero defects.

Keywords: Active learning · Defect detection · Explainable AI · Manufacturing

1 Introduction

Quality control is an integral part of manufacturing systems [1]. Traditionally, it has been a manual process heavily dependent on operators' expertise [2], and often prone to errors. Experience tends to reduce the frequency of such errors, but it is something hard to transfer [3].

Nowadays, Machine and Deep Learning approaches can be used to automate quality control and predict a defect, by identifying abnormalities in process data. This in turn may reduce defects and resources wasted in the processing of defective products [5].

Nevertheless, AI models, becoming increasingly complex, are, in a similar way, increasingly considered as black boxes, difficult to understand and, subsequently, to trust. This requires the introduction of methods to explain their operations and outputs [4].

© The Author(s) 2025
K. Alexopoulos et al. (Eds.): ESAIM 2024, LNME, pp. 40–47, 2025.
https://doi.org/10.1007/978-3-031-86489-6_5

Hence, this paper investigates the combination of an LRP layer to increase the explainability of a CNN model adopted for the identification of defects in a product's 2D images. In addition, a human feedback mechanism is introduced to allow the operators to verify the result of the AI model and label new defect types, thus improving the efficiency of the CNN model over time.

2   Literature Review

AI approaches have been extensively used in manufacturing for scenarios like quality control [6] or predictive maintenance [7]. In [6] a CNN-based algorithm is presented that identifies product defects. Such approaches have also been presented in [8] where the importance of computer vision techniques is explored. Similarly, in [9] a survey on machine and deep learning techniques is conducted aiming to improve and automate the quality control process in manufacturing.

Despite these advancements, and despite the availability of advanced deployment approaches targeting AI systems [10, 11], AI deployment on the shop floor often encounters resistance, primarily due to the complexity of these systems [12]. One of the main reasons is the lack of explainability [13]. While the transparency of AI models and their output is important to ensure human trust in AI, especially those constructed around the use of advanced machine and deep learning techniques, it is often challenging or non-existent [14]. This in turn hinders the application, by shopfloor personnel, of AI-generated results that can improve quality control in real-world environments [15].

Explainable AI (XAI) aims to address these issues by making AI decisions more comprehensible and justifiable. XAI techniques fall into two categories: transparent approaches [16–19], which are inherently interpretable models like decision trees, and post hoc approaches [20, 21], which include both model-specific methods and model-agnostic techniques like SHAP [22] and LIME [23], offering explanations after model training. The potential of XAI to improve quality control is particularly significant in settings that rely heavily on human-AI collaboration. By making AI models more interpretable, XAI allows line operators to understand model uncertainties while enhancing their ability to interact effectively with AI tools [24, 25]. However, the literature indicates a gap in systems that facilitate active feedback from operators to AI models; essential to prevent model drift and to continuously refine AI tools [26].

Active learning is reshaping AI in manufacturing by employing techniques like ensemble methods and contextual bandits, which strategically use informative samples to enhance learning processes [27]. Additionally, uncertainty sampling and selective data querying help AI adapt quickly to new defects and changes in manufacturing, merging operator feedback into training cycles. This fosters a collaborative and adaptive learning environment crucial for achieving zero-defect manufacturing goals [28, 29].

3   Approach

In the study, the primary goal is the introduction of an approach (Fig. 1) that combines advanced AI algorithms, with XAI coupled with a human feedback collection mechanism, to adjust the learning focus of the algorithm dynamically.

A CNN algorithm is selected, due to its versatility and high customizability. The objective of the CNN algorithm is the classification of an image representing a product's surface, based on the presence or absence of a defect. The task of identifying defects is translated into a classification problem due to the simplicity of the image labelling approach in comparison to the complex labelling approach in object detection-related tasks; thus, facilitating the active learning application of the approach.

Subsequently, LRP is applied on top of the CNN model to generate explanations of the outputs of the model. Based on the reviewed literature, and due to the inner complexity of the CNN, LRP is selected. A feedback pipeline is constructed that aims to use the operator's feedback to actively retrain the CNN model.

In more detail, the approach is structured around 8 consecutive steps, which are presented hereafter.

- **Step 1 – CNN model construction**: Images are categorised as defective or non-defective, forming the training dataset for the CNN, which is then deployed for real-time defect detection on the shop floor.
- **Step 2 – LRP rule selection**: After CNN development, the LRP-γ rule is added to introduce an explainable AI layer.
- **Step 3 – Forward pass**: Images pass through the CNN, storing intermediate activations and weights.
- **Step 4 – Relevance initialisation**: Relevance at the output layer is set based on the defect presence classification.
- **Step 5 – Backward relevance propagation**: The relevance score is propagated backwards through the model's network. For each layer of the model, the LRP rule redistributes the relevance from the output layer to the input layer. To achieve this, the LRP-γ rule uses the dimensionless formula detailed in (1).

$$R_j = \sum_i \frac{a_j(w_{ij} + \gamma w_{ij}^+)}{\sum_k a_k(w_{ik} + \gamma w_{ik}^+)} R_i \tag{1}$$

where:

- R_j: The relevance score for a neuron j in the current layer of the model,
- R_i: The relevance score for a neuron i in the next layer of the model,
- a_j: The activation of neuron j in the current layer of the model,
- w_{ij}: The weight connecting neuron j in the current layer to neuron i,
- γ: The Gamma parameter that scales the positive part of the weight,
- $\sum_k a_k(w_{ik} + \gamma w_{ik}^+)$: The normalisation term.

Using the backward propagation results the LRP can generate the necessary explanations which are then provided to the feedback interface, making them easily accessible to an operator. The approach concludes with the following three steps.

- **Step 6 – Operator feedback collection**: Operators use a user-friendly interface to review LRP-generated explanations in real-time, providing feedback (OK or NOK) and identifying potential new defect types. Operators can select between three options (OK, NOK and New class), which is recorded and the classification given by the operators is stored.

- *Step 7 – Feedback collection*: Based on the provided feedback, the dataset is updated in an automated manner weekly. The dataset will be updated to include newly labelled images in existing classes, while new classes are dynamically created based on the users' feedback.
- *Step 8 – Active learning*: Ultimately, CNN is periodically retrained with the updated dataset, improving its predictive accuracy and adapting to new defect types based on the production cycle time.

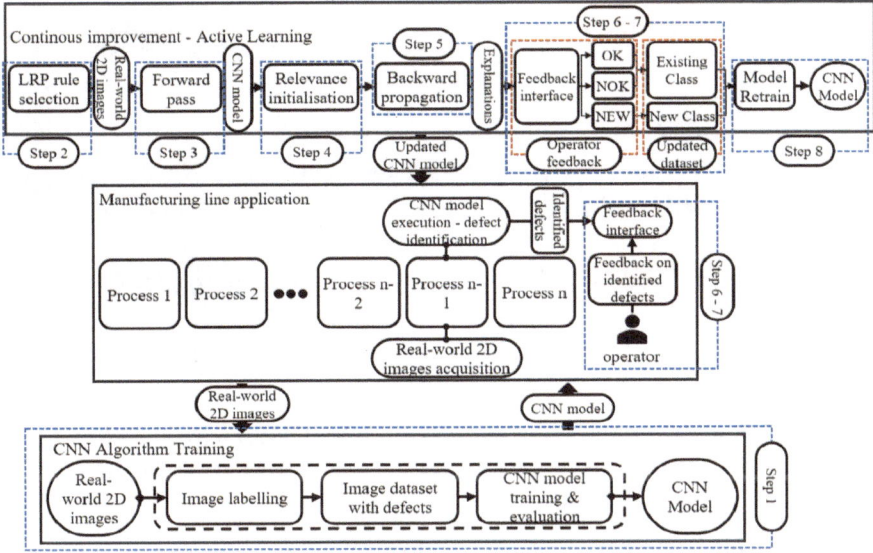

Fig. 1. Active learning approach based on XAI and operator's feedback.

4   Use Case

The proposed approach was applied to a laser welding process in battery module assembly. High-resolution images captured by a 2D vision-based system served as the primary data source, processed on a computing infrastructure consisting of an Intel Core i7-13700H processor and an NVIDIA RTX A1000 GPU. For the initial training of the CNN, a dataset of 1800 mixed real-world and synthetic images was used, enhancing the model's exposure to rare defects. The CNN (Fig. 2), utilized the Adam optimizer, binary cross entropy loss and 10 epochs for better defect detection accuracy.

The CNN model has been trained using the initial 1800 2D images. The CNN model's architecture can be seen in Fig. 2. The performance of the initial CNN model can be seen in Table 1, evaluated using the accuracy, precision, recall and F1-score [30].

LRP was applied to the CNN's outputs to generate explainable visual data, assisting operators in understanding and validating defect detections, as seen in Fig. 3.

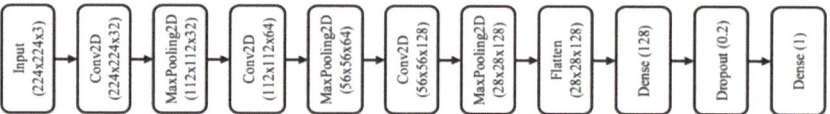

Fig. 2. The architecture of the CNN model.

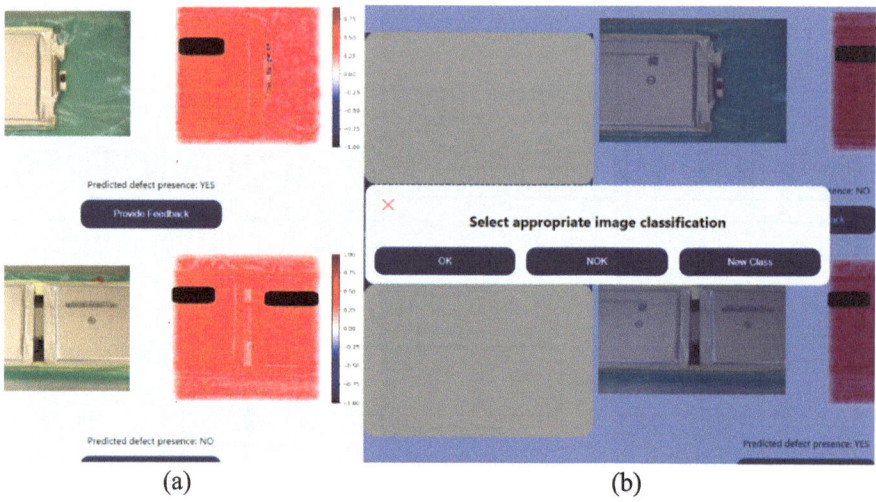

(a)   (b)

Fig. 3. (a) LRP results presented to an operator along with original images and (b) feedback collection interface (grey overlay added on top of real-time data due to data confidentiality).

As seen in Fig. 3, the LRP colorises in red the pixels of an image where defects are not present, while pixels in blue indicate the presence of the class characterising defects. By presenting the original and explained images next to each other, the operator can visually identify defects detected by the CNN. Due to the nature of the battery manufacturing process, defects, such as spots on the weld, can be small, further complicating the manual inspection process. Nevertheless, due to the minute size of the defects, images that contain products with defective welding can potentially be misclassified.

Feedback is collected by operators to actively improve CNN's performance. Feedback provided directly labels the under-examination image with either OK (representing the absence of defect) or NOK (representing the existence of defect). Operators are given a third option to introduce new labels in the dataset in cases where after manual inspection of the under-examination battery a new type of defect is uncovered.

A controlled experiment conducted over a week gathered substantial feedback, leading to the retraining of the CNN, whose performance can be seen in Table 1.

Post-retraining, improvements were observed: accuracy increased marginally from 0.89 to 0.9, while recall improved from 0.57 to 0.66, and the F1-score from 0.70 to 0.77, indicating enhanced reliability and reduced risk of missing defects.

The results presented in Table 1 underline the importance of enabling operators to provide feedback on data they utilise daily to ensure that quality standards are met. Given

Table 1. Performance comparison of original and retrained CNN model.

Metric	Original CNN	Retrained CNN
Accuracy	0.89	0.9
Precision	0.94	0.94
Recall	0.57	0.66
F1-score	0.70	0.77

the rapid digitalisation of today's manufacturing, AI systems are adopted and operators are faced with the challenge of understanding the results produced by them. With the application of XAI, operators demonstrated an increased reliance on the outputs of the CNN, given the explanations the LRP layer provided them. On top of this, the integration of humans in the loop of continuous model improvement is an effective approach given the increase in the CNN model's performance; thus, validating the effectiveness of integrating active learning and XAI in manufacturing environments.

Lastly, operators were asked to provide feedback on the usefulness of introducing explainability to the AI system. Approximately 83% of operators (7 men and 5 women) deemed the introduction of explainability as useful since it allows them to become part of the constant improvement process and speed up the manual defect identification phase, while the rest pointed out that improvements should be made to the feedback interface to increase its user-friendliness, such as removing pop-up dialogues and incorporating the feedback options beneath the model's output.

5   Conclusion

This study presents an approach for detecting subtle defects in battery module assembly lines, utilizing a semi-supervised CNN for initial defect detection. LRP is then applied to the CNN's outputs to visually explain the decision-making process, pinpointing why specific areas are flagged as defective, enhancing transparency and supporting the validation of detected defects, thus, providing clarity in quality control processes.

Operators evaluate areas highlighted by LRP to confirm defect identifications and to uncover new types of defects. This is essential for periodic retraining of the CNN, enabling the system to adapt to evolving manufacturing conditions and continuously refine its accuracy. This dynamic approach has been tested in a battery modules assembly use case with its result demonstrating the potential to improve defect detection but also to establish a robust mechanism for enhancement of quality control practices.

Future research will be focused on further optimising the active learning pipeline and expanding the areas of application of the proposed approach through the adoption of the LRP layer on more diverse manufacturing scenarios. Also, future research will aim to extend the approach to account for the uncertainty of defects in the process of feedback collection by operators to assist them during the feedback collection phase.

Acknowledgement. This work was partially supported by the HORIZON-CL4-2021-TWIN-TRANSITION-01 openZDM project, under Grant Agreement No. 101058673.

References

1. Chryssolouris, G.: Manufacturing Systems: Theory and Practice. Springer, New York (2006)
2. Medici, V., et al.: Integration of non-destructive inspection (NDI) systems for zero-defect manufacturing in the industry 4.0 era. In: 2023 IEEE International Workshop on Metrology for Industry 4.0 & IoT (MetroInd4.0&IoT), pp. 439–444. IEEE, Brescia, Italy (2023)
3. Torres, Y., Nadeau, S., Landau, K.: Classification and quantification of human error in manufacturing: a case study in complex manual assembly. Appl. Sci. **11**, 749 (2021)
4. Ntoulmperis, M., et al.: 3D point cloud analysis for surface quality inspection: a steel parts use case. Procedia CIRP **122**, 509–514 (2024)
5. Jia, Z., Wang, M., Zhao, S.: A review of deep learning-based approaches for defect detection in smart manufacturing. J. Opt. **53**, 1345–1351 (2024)
6. Sundaram, S., Zeid, A.: Artificial intelligence-based smart quality inspection for manufacturing. Micromachines **14**, 570 (2023)
7. Cerquitelli, T., Nikolakis, N., O'Mahony, N., Macii, E., Ippolito, M., Makris, S. (eds.): Predictive Maintenance in Smart Factories: Architectures, Methodologies, and Use-cases. Springer Singapore, Singapore (2021)
8. Andrianandrianina Johanesa, T.V., Equeter, L., Mahmoudi, S.A.: Survey on AI applications for product quality control and predictive maintenance in industry 4.0. Electronics **13**, 976 (2024)
9. Mueller, C., Mezhuyev, V.: AI models and methods in automotive manufacturing: a systematic literature review. In: Al-Emran, M., Shaalan, K. (eds.) Recent Innovations in Artificial Intelligence and Smart Applications, pp. 1–25. Springer International Publishing, Cham (2022)
10. Nikolakis, N., et al.: A microservice architecture for predictive analytics in manufacturing. Procedia Manuf. **51**, 1091–1097 (2020)
11. Bowden, D., et al.: A cloud-to-edge architecture for predictive analytics
12. Gabsi, A.E.H.: Integrating artificial intelligence in industry 4.0: insights, challenges, and future prospects–a literature review. Ann. Oper. Res. (2024)
13. Felderer, M., Ramler, R.: Quality assurance for AI-based systems: overview and challenges (introduction to interactive session). In: Winkler, D., Biffl, S., Mendez, D., Wimmer, M., Bergsmann, J. (eds.) Software Quality: Future Perspectives on Software Engineering Quality, pp. 33–42. Springer International Publishing, Cham (2021)
14. Walmsley, J.: Artificial intelligence and the value of transparency. AI Soc. **36**, 585–595 (2021)
15. Wang, X., Yang, Y., Tao, D., Zhang, T.: The impact of AI transparency and reliability on human-AI collaborative decision-making. In: Presented at the AHFE 2023 Hawaii Edition (2023)
16. Gohel, P., Singh, P., Mohanty, M.: Explainable AI: current status and future directions
17. Letzgus, S., Wagner, P., Lederer, J., Samek, W., Muller, K.-R., Montavon, G.: Toward explainable artificial intelligence for regression models: a methodological perspective. IEEE Signal Process. Mag. **39**, 40–58 (2022)
18. Saeed, W., Omlin, C.: Explainable AI (XAI): a systematic meta-survey of current challenges and future opportunities. Knowl. Based Syst. **263**, 110273 (2023)
19. Janzing, D., Minorics, L., Blobaum, P.: Feature relevance quantification in explainable AI: a causal problem
20. Poché, A., Hervier, L., Bakkay, M.-C.: Natural example-based explainability: a survey. In: Longo, L. (ed.) Explainable Artificial Intelligence, pp. 24–47. Springer Nature Switzerland, Cham (2023)
21. Gianfagna, L., Di Cecco, A.: Model-agnostic methods for XAI. In: Explainable AI with Python, pp. 81–113. Springer International Publishing, Cham (2021)

22. Zhang, K., Xu, P., Zhang, J.: Explainable AI in deep reinforcement learning models: a SHAP method applied in power system emergency control. In: 2020 IEEE 4th Conference on Energy Internet and Energy System Integration (EI2), pp. 711–716. IEEE, Wuhan, China (2020)
23. Nguyen, H.T.T., Cao, H.Q.: Evaluation of Explainable Artificial Intelligence: SHAP, LIME, and CAM
24. Müller, R., Reindel, D.F., Stadtfeld, Y.D.: The benefits and costs of explainable artificial intelligence in visual quality control: evidence from fault detection performance and eye movements. Hum Ftrs Erg Mfg Svc. (2024)
25. Alexander, Z., Chau, D.H., Saldaña, C.: An interrogative survey of explainable AI in manufacturing. IEEE Trans. Ind. Inf. **20**, 7069–7081 (2024)
26. Wang, Z., Liu, Y., Thiruselvi, A.A., Hamou-Lhadj, A.: XAIport: A Service Framework for the Early Adoption of XAI in AI Model Development (2024)
27. Zeng, Y., Chen, X., Jin, R.: Ensemble active learning by contextual bandits for AI incubation in manufacturing. ACM Trans. Intell. Syst. Technol. **15**, 1–26 (2024)
28. Song, Y., Li, F., Wang, Z., Zhang, B., Zhang, B.: Uncertainty quantification of data-driven quality prediction model for realizing the active sampling inspection of mechanical properties in steel production. Int. J. Comput. Intell. Syst. **17**, 74 (2024)
29. Johnson, M., Albizri, A., Harfouche, A., Fosso-Wamba, S.: Integrating human knowledge into artificial intelligence for complex and ill-structured problems: informed artificial intelligence. Int. J. Inf. Manage. **64**, 102479 (2022)
30. Powers, D.: Evaluation: from precision, recall and F-factor to ROC, informedness, markedness & correlation. Mach. Learn. Technol. **2** (2008)

Synthetic Data for AI-Powered Ultrafast Laser Based Micro-structuring Method Description

Beatriz Blanco-Filgueira[1]($\boxtimes$), Tamara Delgado[1], Andrea Gregores Coto[1], Céline Petit[2], David Bruneel[2], Pablo Romero[1], and Santiago Muiños-Landin[1]

[1] AIMEN Technology Centre, 36410 Porrino, Pontevedra, Spain
beatriz.blanco@aimen.es
[2] LASEA, 4102 Seraing, Belgium

Abstract. High-power ultrashort pulse lasers (USPLs) represent a sustainable alternative for industrial surface processing. Laser Surface Texturing (LST) allows modification of surface properties such as mechanical, chemical or optical properties. The adoption of USPLs in industry will enable the surface functionalization of large 3D parts through the LST process. However, a real exploitation of USPLs still faces limitations related to high processing resolution, quality, productivity, and the need for expert knowledge in laser micro-structuring strategies. It is crucial to increase maturity, reliability and throughput of LST on a large scale, as well as to provide the industry with the necessary tools for selecting the suitable laser process for each specific application.

Achieving these goals requires significant effort in data collection, which is costly in terms of processing times, results analysis and human resources. Additionally, the availability of software simulators for laser micro-structuring is limited, posing challenges for describing USPLs processes through numerical models and methods. In this context, Artificial Intelligence (AI) models represent a unique tool to enhance process understanding and results analysis, providing rapid prediction and visualization capabilities. However, developing AI-models requires vast amounts of data, which are not easily obtainable experimentally, but simulators can play a vital role in this aspect.

We present a combined solution based on machine learning methods fed by synthetic data generated by LS-Plume® from LASEA for predicting femtosecond laser single beam-based grooves on stainless steel 316L. The methodology for developing such applications is described, and several regression models are compared. The analysis of metrics and prediction error demonstrates that leveraging the synergy between simulation tools and AI-models can be an effective strategy for training AI-models, thereby avoiding the need for a complete, systematic, experimental parameter sweep approach. Consequently, the improvement of numerical models is a valuable strategy for training AI-models that can boost the adoption of USPLs for laser micro-structuring on a large industrial scale.

© The Author(s) 2025
K. Alexopoulos et al. (Eds.): ESAIM 2024, LNME, pp. 48–59, 2025.
https://doi.org/10.1007/978-3-031-86489-6_6

Keywords: laser micro-structuring · surface texturing · ultrashort pulse laser · light-matter interaction · synthetic data · AI-models

1 Introduction

Ultrashort pulse lasers (USPLs) emit pulses of light with a duration in the femtosecond (fs) or picosecond (ps) range, enabling the fabrication of high-precision structures at the micro- and nanoscale. Furthermore, USPLs facilitate superior quality in laser micromachining processes, with negligible thermal effects [13,14,16]. Recent developments in USPLs technology have resulted in high-power and high-pulse repetition rate (tens of MHz) laser systems, bringing USPLs closer to meeting the industrial demands not only in terms of processing quality but also productivity. Furthermore, efforts to develop high-speed beam scanning devices and multi-beam processing methods are also been made to achieve the necessary processing speeds for industrial applications with USPLs [5,18]. Another critical aspect for scaling up USPL laser processing to industrial scale is the required knowledge in laser processing for each specific application, such as laser drilling, laser cutting, or laser surface texturing.

Laser surface texturing (LST) is used to modify mechanical, chemical, electrical, or optical properties of a material by engraving a texture on its surface. LST has several applications, including wettability control, friction improvement, and increased adhesive properties, among others [11,17,19]. For the effective industrial exploitation of LST with USPLs, it is necessary to provide the industry with the tools required to select the optimal laser processing strategy to achieve the desired surface functionality.

Optimal LST process parameters can be found using gradient search method, where one parameter is changed at a time. Alternatively, the process can be modeled, but incorporating subtle effects and complexities is challenging and can become computationally intractable. Artificial Intelligence (AI) methods emerge as an promising alternative, though their potential for predictive visualization of USPLs LST has not been extensively explored. For instance, in [6] a Neural Network was used to predict the surface profile based on binary images from a digital micromirror device, although it was limited to fixed laser parameter values. Another study used a Convolutional Neural Network (CNN) trained with experimental images to predict and optimize dimple depth and crown height on grey cast iron, [12]. Recent research used regression models to correlate laser system parameters with dimple diameter, [7]. In a different study, machine learning (ML) techniques were proposed for predicting and optimizing femtosecond laser percussion drilling in a nickel-based single-crystal superalloy, [20]. Another example combined theoretical modeling and ML to predict the surface width and depth of tapered microchannels in silica glass, [10].

State-of-the-art research demonstrates a high interest in leveraging ML techniques to optimize various UPSL precision manufacturing processes. However, the diversity of applications and materials requires the use of different modeling methods and tailored solutions, making further efforts to demonstrate ML capabilities and effective methodologies for prediction and optimization crucial. In this context, simulation tools can play a new role as trusted synthetic data generators to feed AI-models. Rather than being replaced by ML methods, software tools can be a key component in the early stages of their development due to the high cost of experimental data collection in terms of human resources, processing times, and analysis. Moreover, whereas the former may be limited to prediction capabilities, often with a high computational cost, AI-models offer faster prediction and process optimization capabilities.

In this work, a method description for AI prediction of ultrafast laser based micro-structuring using synthetic data is presented. Specifically, several ML regression models are developed and compared for predicting femtosecond laser single beam-based grooves on stainless steel 316L. The features of the models include process parameters such as pulse repetition rate, scanning speed, pulse energy, beam diameter, number of passes and pattern pitch. Synthetic data from LASEA LS-Plume® simulation software were used to guide the development of AI solutions for these procedures. The results for metrics (MAE, MSE, RMSE and R^2) along with the prediction error of groove's depth and Full Width at Half Maximum (FWHM), provide further evidence that AI-methods significantly aid in meeting the needs of this field.

2   Methods

2.1   Synthetic Data Generation: LASEA LS-Plume®

Synthetic data were generated by using the simulator LS-Plume® from Lasea [9]. This tool allows the simulation of ablation surface topographies due to ultrafast laser processing for three machining strategies: scanned line or area, precession, and percussion. The use of several common materials like copper, titanium, and stainless steel 316L is also available. LS-Plume® can predict the ablation profile based on input parameters classified into two categories: laser parameters and material properties. The laser parameters take into account all the user-defined parameters: pulse energy, pulse repetition rate, spot size at focus, and scanning parameters (speed, pitch between two consecutives engraved lines, and incident angle) that can be chosen during a machining. The material parameters are the initial surface topography, the ablation threshold, the laser penetration depth, and the complex refractive index. The engineering model used to obtain ablation profiles for given processing parameters and material considers several effects appearing during the laser-matter interaction like incubation, surface reflectivity, and Gaussian beam diameter variation with the distance to the focal plane. More details about the model can be found here [2].

The LS-Plume® has been developed to respond quickly to industrial needs. In this context, the purpose of the simulator is to predict ablation profile thanks to a simple model. Since thermal effects are not modeled, the performances of the simulator, good match between simulated profiles and experimental profiles, are maximum in a regime where thermal effects during ablation are not significant.

The simulated data were obtained by using the scanned line strategy where an ablation profile is simulated by the overlapping of parallel lines space at an identical pitch. As explained in Sect. 2.2, different combinations of laser parameters were tested for stainless steel 316L by varying the physical quantities that can be chosen by the software user. Given the amount of data generated, an automated data generation procedure was developed using Python code.

2.2 Dataset Description

Training and testing any machine learning model demands a considerable amount of data with good coverage in terms of the process features values. For this purpose, simulations of single beam-based patterns (DW) at 100 kHz repetition rate and 300 fs pulse width were performed, varying scanning speed, pulse energy, beam diameter, number of passes and pattern pitch from 10 to 1000 mm/s, 2.87 to 144.71 μJ, 10 to 50 μm, 1 to 10, and 125 to 300% of the beam diameter, respectively. This results in a total of 3326 groove patterns based on different combinations of the features values. The dataset was split into train and test datasets, including 2827 and 499 patterns, which means that %15 of the whole dataset was held for the test. Shuffling was applied to the data before the split and features were standardized to zero mean and unit variance using the mean and standard deviation of the training samples.

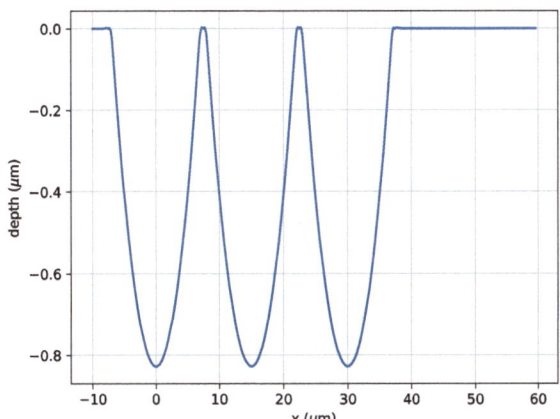

Fig. 1. Example of a simulated curve profile of a single beam-based groove pattern.

Each pattern is represented by a profile curve as depicted in Fig. 1, taken as an example. Depth and FWHM (Full Width at Half Maximum) are among the most representative characteristics of the groove. Both are calculated from each curve profile and used as targets for the machine learning models. FWHM prediction was performed in a reduced dataset, removing those patterns with a pitch less than 1.5 times of the beam diameter. The reason is that for these configurations the adjacent grooves overlap until they join and therefore the width of a single groove cannot be defined.

Finally, data augmentation was applied in order to complete the dataset and to improve its generalization ability and accuracy. In addition to the profile curves generated from the simulation data, zero-energy data were artificially created and added to the training dataset, where the input energy value was zero, and also the depth and FWHM. The same procedure was done for zero number of passes. This augmentation was based on the physical assumption that there is no groove when there is no incident pulse. Once augmented, the training datasets consisted of 2928 and 2374 patterns for depth and FWHM prediction, respectively, which correspond to 3427 and 2776 patterns full datasets.

2.3   Machine Learning Models for Non-linear Regression

The aim of this work is to provide an initial prediction of the LST result to the expert for a given configuration of the set-up. It was tackled as a supervised machine learning problem, training models on a dataset that correlates the features of the process with the targets that describe the profile curve.

The complex relationships between features and target variables require the use of non-linear regression algorithms, machine learning techniques able to capture the complexity of light-matter interaction that can not be effectively represented by a linear model. For this purpose, the performance of a variety of some popular non-linear regression algorithms was compared to address the problem.

The models were implemented by using different classes from scikit-learn library. A cross-validation grid-search over a parameter grid for hyperparameter optimization was applied for each model using the GridSearchCV method. It consisted on an exhaustive search over the main model's parameters using a 5-fold cross-validation strategy. Once the search is finished, the model is refit using the best found parameters on the whole training dataset. The Mean Square Error (MSE) was used as score or training loss, and different wide-spread metrics for the test: MAE, MSE, RMSE and R^2.

The models and their optimized hyperparameters are briefly described below, assigning an acronym to each of them as a reference for the tables and figures in Sect. 3. Firstly, a decision tree regressor was developed by using DecisionTreeRegressor class ("dtr") and optimizing the maximum depth of the tree. Decision trees are simple and explainable but can be unstable, what can be mitigated by the use of a random forest regressor. Thus, RandomForestRegressor class ("rfr") was used optimizing the maximum depth of the tree but also the number of trees in the forest.

Four algorithms that work well with small datasets and avoid overfitting to some extent were also tried. Gradient boosting, GradientBoostingRegressor class ("gbr"), which builds an additive model in a forward stage-wise fashion, fitting a regression tree in each stage on the negative gradient of the given loss function. The optimized hyperparameters are the maximum depth of the individual regression estimators, the number of boosting stages to perform, the minimum number of samples required to split an internal node and to be at a leaf node, and the number of features to consider when looking for the best split. Two state of the art regressors, XGBoost ("Extreme Gradient Boosting" [3], "xgb"), an efficient and scalable implementation of gradient boosting framework [4], and LightGBM [8] ("lgbm"), that includes gradient-based one-side sampling (GOSS) and exclusive features bundling (EFB), were implemented and optimized in terms of the maximum depth of a tree, the number of boosted trees to fit, and the learning rate. Finally, another gradient boosted decision tree implementation, CatBoost [15] ("cbr"), was used optimizing the same hyperparameters as the later and also the coefficient at the L2 regularization term of the cost function.

Other classical approaches were tested as well. A k-nearest neighbors based regressor using the KNeighborsRegressor class ("knn"), predicting by local interpolation of the targets associated of the nearest neighbors in the training set and optimizing the default number of neighbors to use, and an epsilon-support vector regression by the use of SVR class ("svr"), optimizing the regularization parameter and the distance epsilon from the actual value within which no penalty is associated in the training loss function. Finally, MLPRegressor class was used to develop a multi-layer perceptron regressor ("mlp"), based on Adam solver, using ReLU activation function, and shuffling samples in each iteration. Different configurations for the number and size of the hidden layers were studied, optimizing the strength of the L2 regularization term, the size of minibatches, the initial learning rate and the maximum number of epochs.

Each model was trained for both depth and FWHM prediction and then MultiOutputRegressor class was used to extend the models for multi target prediction, training separate and independent models, one per each target variable. Decision tree and k-nearest neighbors based regressors are also able to handle multi-output problems.

3   Results and Discussion

Evaluation of all models for single and multi-target prediction was performed. In all cases, the learning curve revealed that the number of samples in the training dataset allow a good generalization. Moreover, the influence of every hyperparameter on the training and test score was studied to ensure that the model does not suffer from overfitting or underfitting for the optimized values.

Table 1. Comparison of machine learning models prediction performance on test dataset.

Model	depth				FWHM			
	MAE	MSE	RMSE	R^2	MAE	MSE	RMSE	R^2
dtr (Decision tree)	0.14	0.11	0.33	0.9997	0.17	0.10	0.32	0.9989
rfr (Random forest)	0.24	0.35	0.59	0.9992	0.18	0.09	0.29	0.9999
gbr (Gradient boosting)	0.14	0.11	0.33	0.9997	0.11	0.04	0.20	0.9996
xgb (Extreme GB)	0.14	0.12	0.35	0.9997	0.11	0.04	0.20	0.9996
lgbm (LightGBM)	1.99	12.37	3.52	0.9709	0.30	0.17	0.42	0.9981
cbr (CatBoost)	0.13	0.09	0.29	0.9998	0.11	0.03	0.18	0.9996
svr (Support vector)	3.86	28.00	5.29	0.9340	1.65	4.18	2.04	0.9551
knn (K-neighbors)	3.04	81.33	9.02	0.8083	1.29	4.76	2.18	0.9489
mlp (Perceptron)	0.62	2.14	1.46	0.9946	0.31	0.17	0.42	0.9982

Table 2. Comparison of machine learning models performance: depth and FWHM prediction using MultiOutputRegressor class (left) and multi-target models (right) on test dataset.

Model	MAE	MSE	RMSE	R^2
dtr	0.19	0.34	0.54	0.9988
rfr	0.31	0.80	0.76	0.9980
gbr	0.18	0.29	0.46	0.9992

Model	MAE	MSE	RMSE	R^2
dtr	0.23	0.30	0.54	0.9986
knn	2.71	46.82	6.08	0.8613

The metrics introduced in Sect. 2.3 were used to compare the models. The results are presented in Table 1 for depth and FWHM prediction on unseen test dataset. Similarly, some examples of MultiOutputRegressor class and native multi-target models potential for both targets prediction is shown in Table 2. They illustrate that these estimators are good predictors of the main characteristics of the grooves engraved by DW LST.

The regressors performance described by metrics in Table 1 is visualized in Fig. 2 and Fig. 3, depicting the prediction error of every model for depth and FWHM, that is, the difference between the actual and predicted values over the test dataset.

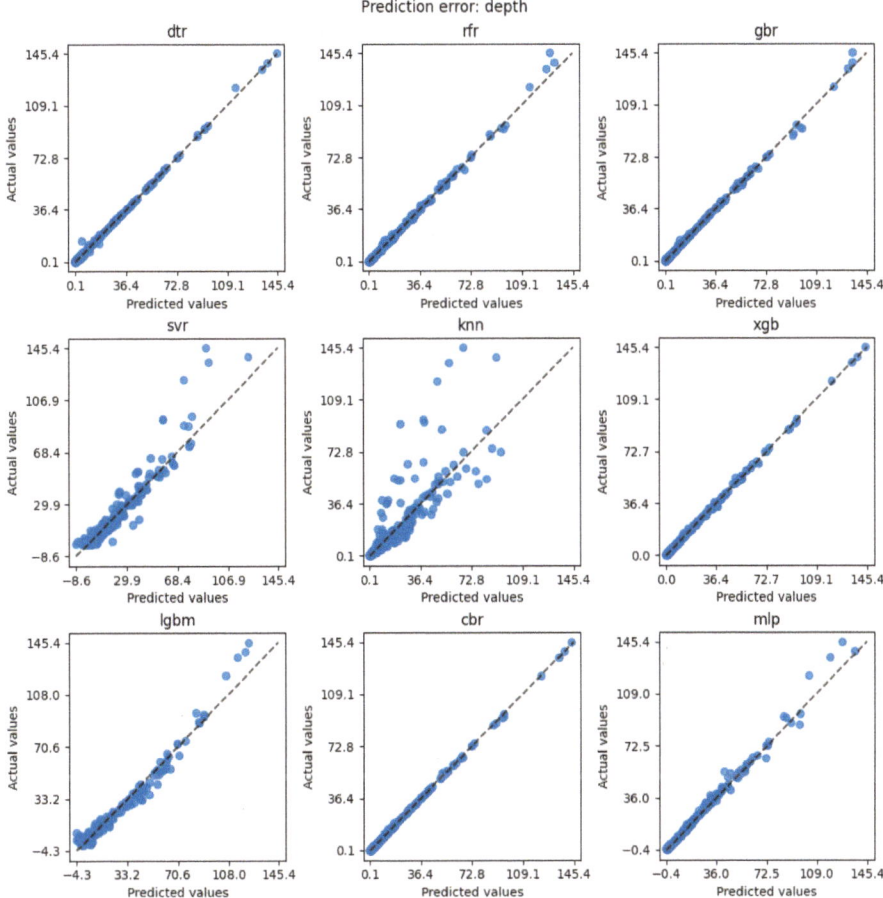

Fig. 2. Prediction error of the depth.

The features importance is of interest to verify the consistency of the models as well. It could be also of great worth to an expert when configuring a set-up. In scikit-learn, the importance of a feature is computed as the normalized total reduction of criteria by feature, what is sometimes called Gini importance or

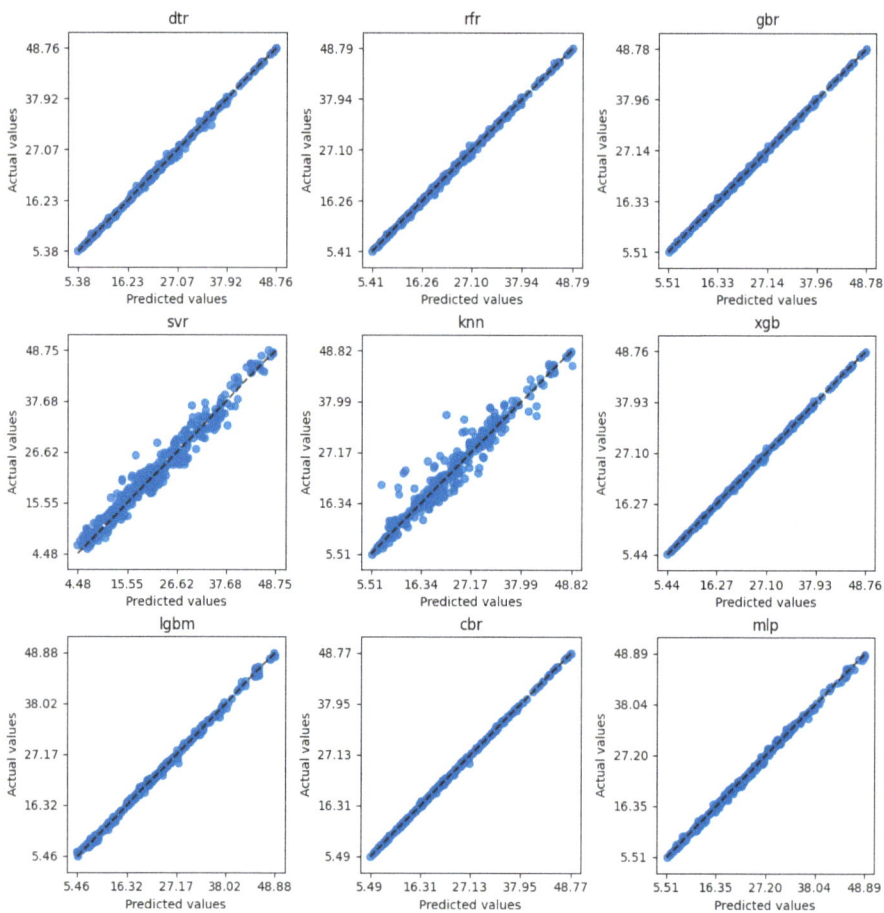

Fig. 3. Prediction error of the FWHM.

"mean decrease impurity" [1]. Figure 4 and Fig. 5 show the features importance in depth and FWHM predictions for the models which have this attribute. As expected, considering that moderate values of the energy were taken, most of the models agree that the most relevant feature for the depth is the scanning speed, followed by the number of passes. On the other hand, the beam diameter and the pulse energy are more decisive to define the FWHM. When considering both targets, the scanning speed and the number of passes are the most representative.

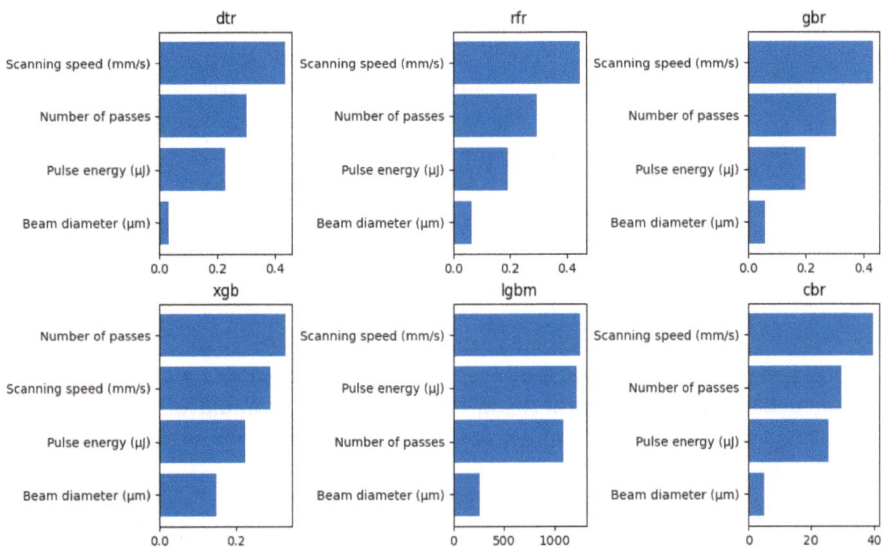

Fig. 4. Features importance for depth prediction.

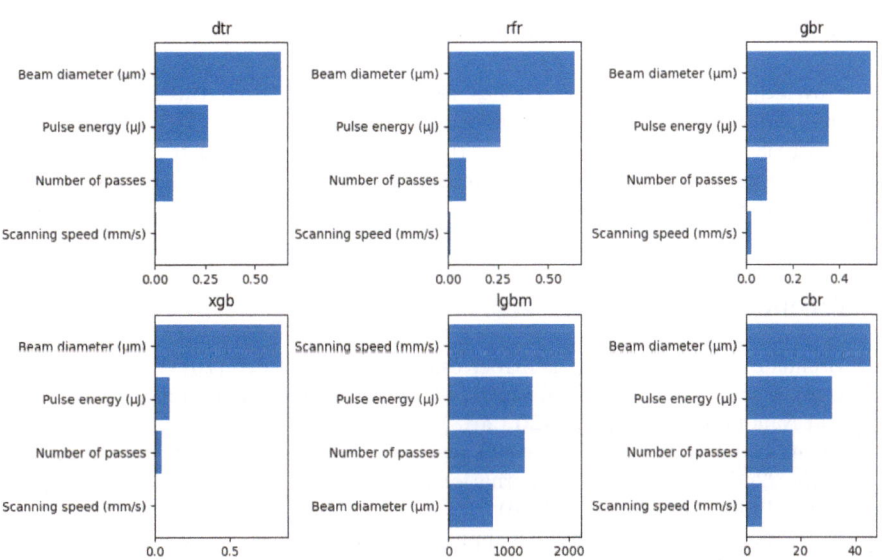

Fig. 5. Features importance for FWHM prediction.

4   Conclusions and Future Work

Synthetic data for single beam-based groove curve profile were generated through the simulator LS-Plume® from LASEA. These data were used to fed several ML regression models for predicting the depth and Full Width at Half Maximum (FWHM) of grooves on stainless steel 316L. The process parameters used as models features were pulse repetition rate, scanning speed, pulse energy, beam diameter, number of passes and pattern pitch. Data augmentation for zero-energy and zero number of passes was applied to complete the dataset and improve generalization. A comprehensive methodology for developing machine learning models for predicting the groove's profile curve for a given configuration of the set-up was presented.

The performance of a variety of popular non-linear regression algorithms was evaluated and compared using metrics such as MAE, MSE, RMSE and R^2, demonstrating promising prediction error results. This study serves as a useful guide for the development of AI-models for predicting ultrafast laser surface texturing (USPL LST) processes, highlighting the capabilities of ML methods and the benefits of using synthetic data generated from simulators. Additionally, multi-target prediction was analysed, which is particular relevant for extending this work towards process optimization.

Future work includes ongoing experimental validation of the models, with a dataset currently in progress. Generative AI could be employed to feed the models during the training and refining processes. Furthermore, the authors are also considering exploring the integration of physics-informed models to improve prediction accuracy and reliability. Finally, the use of these tools for process optimization will be also studied.

Acknowledgment. This research has received funding from the Horizon Europe research and innovation programme under the project OPERATIC with Grant Agreement 101058409. The authors want to thank the comments and fruitful discussions with all the members of the Smart Systems and Smart Manufacturing (S3M) and the Advanced Manufacturing Processes departments of the AIMEN Technology Centre.

References

1. Breiman, L.: Classification and regression trees. Routledge (2017)
2. Cangueiro, L., et al.: Model for ultrafast laser micromachining. Laser-based Micro- and Nanoprocessing XII **10520** (2018)
3. Chen, T., Guestrin, C.: XGBoost: a scalable tree boosting system. In: Proceedings of the 22nd ACM SIGKDD International Conference on Knowledge Discovery and Data Mining, pp. 785–794 (2016)
4. Friedman, J.H.: Greedy function approximation: a gradient boosting machine. Ann. stat., 1189–1232 (2001)
5. Gillner, A., Finger, J., Gretzki, P., Niessen, M., Bartels, T., Reininghaus, M.: High power laser processing with ultrafast and multi-parallel beams. J. Laser Micro/Nanoengineering **14**(2) (2019)
6. Heath, D.J., et al.: Machine learning for 3D simulated visualization of laser machining. Opt. Express **26**(17), 21574–21584 (2018)
7. Hoskins, J.K., Hu, H., Zou, M.: Exploring machine learning and machine vision in femtosecond laser machining. ASME Open J. Eng. **2** (2023)
8. Ke, G., et al.: LightGBM: a highly efficient gradient boosting decision tree. Adv. Neural Inf. Process. Syst. **30** (2017)

9. LASEA: Homepage - ls-plume. https://www.ls-plume.com/ (2020). Accessed 8 May (2024)
10. Liao, K., et al.: Shape regulation of tapered microchannels in silica glass ablated by femtosecond laser with theoretical modeling and machine learning. J. Intell. Manuf. **34**(7), 2907–2924 (2023)
11. Mandolfino, C., Obeidi, M., Lertora, E., Brabazon, D.: Comparing the adhesion strength of 316L stainless steel joints after laser surface texturing by CO_2 and fiber lasers. Int. J. Adv. Manuf. Technol. **109**(3), 1059–1069 (2020)
12. McDonnell, M.D., et al.: Machine learning for multi-dimensional optimisation and predictive visualisation of laser machining. J. Intell. Manuf. **32**, 1471–1483 (2021)
13. Nishizawa, N.: Ultrashort pulse fiber lasers and their applications. Jpn. J. Appl. Phys. **53**(9), 090101 (2014)
14. Nolte, S., Schrempel, F., Dausinger, F.: Ultrashort pulse laser technology. Springer Ser. Opt. Sci. **195**(200), 1 (2016)
15. Prokhorenkova, L., Gusev, G., Vorobev, A., Dorogush, A.V., Gulin, A.: CatBoost: unbiased boosting with categorical features. In: Advances in Neural Information Processing Systems, vol. 31 (2018)
16. Račiukaitis, G.: Ultra-short pulse lasers for microfabrication: a review. IEEE J. Sel. Top. Quantum Electron. **27**(6), 1–12 (2021)
17. Riveiro, A., et al.: Laser texturing to control the wettability of materials. Procedia CIRP **94**, 879–884 (2020)
18. Schille, J., Schneider, L., Streek, A., Kloetzer, S., Loeschner, U.: High-throughput machining using a high-average power ultrashort pulse laser and high-speed polygon scanner. Opt. Eng. **55**(9), 096109–096109 (2016)
19. Shamsul Baharin, A.F., Ghazali, M.J., A Wahab, J.: Laser surface texturing and its contribution to friction and wear reduction: a brief review. Ind. Lubrication Tribol. **68**(1), 57–66 (2016)
20. Zhao, Z., et al.: Design of a femtosecond laser percussion drilling process for Ni-based superalloys based on machine learning and the genetic algorithm. Micromachines **14**(11), 2110 (2023)

Towards Explicable AI in Systemic Identification of Surrogate Models of Manufacturing Processes

Alexios Papacharalampopoulos, Christos Papaioannou, Olga Maria Karagianni, and Panagiotis Stavropoulos[✉]

Laboratory for Manufacturing Systems and Automation, Mechanical Engineering and Aeronautics Department, University of Patras, 26504 Patras, Greece
pstavr@lms.mech.upatras.gr

Abstract. Surrogate models of manufacturing processes are highly useful in the context of digital twins, as they can be considered as a prerequisite for linking between physics and real machines' cases. However, in many cases of surrogate models, the so-called (hyper)parameters are not easy to be estimated. In this work, the role of AI is investigated in terms of its efficiency in doing that in an (semi) automated way. The case of ARX models is considered for this, where the parameters are clearly related to the physics of the process. In particular, the technique of systemic identification is adopted. The aforementioned investigation is performed in this particular case regarding intuitiveness with respect to the physics, through adopting AI techniques that could be considered to be explicable in some sense. The results indicate the limitations of the AI techniques and their link to the process dynamics as well as their relationships with traditional techniques.

Keywords: Manufacturing Process · Digital Twin · Surrogate model

1 Introduction

Artificial Intelligence (AI) integration in manufacturing is an emerging phenomenon, concurrently occurring simultaneously with the inclusion of human [1]. This is in line with Industry 5.0, since human centricity but also integration of Industry 4.0 Key enabling technologies are in place. As such, under the context of AI, there are various types of digital twins and relevant operations [2], especially at manufacturing process level, aiming real-time monitoring and optimization. In fact, the Digital Twin, being an "umbrella" of various technologies, is subject to architecture design and its structure [3] can vary, including many different modules, both in terms of hardware and software (integrating also the concept of Cyber-Physical Systems).

In particular, surrogate models are highly useful in forming a digital twin [3], as they are able to run in near-real-time (as opposed to simulations) and provide feedback about the status of the manufacturing process as well they can generate guidelines/control. They can be either data driven (utilizing techniques such as machine learning [4] and regression [5]), or deriving from simulations, i.e. numerical methods, after their acceleration (utilizing techniques such as ROMs [6, 7]).

© The Author(s) 2025
K. Alexopoulos et al. (Eds.): ESAIM 2024, LNME, pp. 60–68, 2025.
https://doi.org/10.1007/978-3-031-86489-6_7

To achieve designing and implementing a digital twin and/or a surrogate model, one has to face the concept of identification [8, 9]. Its significance is high in those procedures, as the equations of the models behind the Digital Twins are taking their final form. Such parameters' identification involves the estimation of the order of Auto-Regressive (ARX) models, or even the meta-parameters [10] engaged in the case of Machine Learning. The two-fold significance of Neural Networks (NN) in such a context has to with two facts: firstly, NN's can be used as models themselves, but, secondly, they can be used as estimators of the order of ARX models. In the context of explicable AI as a means of human-centric technology, the added value of the current work is that it constitutes a numerical investigation of the **capabilities of the NN's** to undertake the aforementioned roles. Also, the NN's themselves can give information on why particular ARX models ought to be used. This is highly relevant, since automating to an extent the choice of the model and/or the hyper-parameters (i.e. herein the order of the ARX model) is a major challenge. This is evident in literature [11–13], even in manufacturing, i.e. in toolwear detection [14] or laser processes quality monitoring [15].

This work is the first step in a long chain of research steps and works under the assumption of linear systemic responses, i.e. being a weighted sum of exponential/sinusoidal functions and aims at providing **intuition** to human operators of digital twins. The exact benefit of the current work is the proof of concept that, for simple cases, order identification can potentially be automated in an intuitive way.

2 Method

The method constitutes mainly of numerical investigations. As briefly presented in Fig. 1, the dataset consists of a large number of responses, which have been prepared in a way so that they are distinguishable from one another up to an extent. More specifically, responses of linear systems such as $\sum_{n=1}^{Q} A_n e^{\lambda_n t} \sin(\omega_n t)$ are considered, with Q being the order of the system that has this response, A_n being numerically significant amplitudes and λ_n, ω_n being distinguishable time constants. This type of responses is derived from linear differential equations, or their discrete equivalent [16], the ARX models (auto-regressive models with exogenous input).

The involved mathematical assumptions are:

(I) response has to have time constants larger than sampling period and less than sampling interval
(II) there are adequate numerical differences between the constituents of the responses (i.e. the amplitudes contribute significantly)
(III) overfitting probability is reduced

The last one is guaranteed through a simple estimation: the dataset sample is at minimum three times larger (per class) than the number of the coefficients that the NN has. The various estimators are then investigated with respect to their efficiency in predicting the order of the response in MATLAB, after a training phase. The dataset itself consists of various classes, depending on the type of the responses: (a) first order responses are of $A_1 e^{\lambda_1 t}$ type, (b) second order responses are of four different types, depending on the poles placement: $A_1 e^{\lambda_1 t} \sin(\omega_1 t + \varphi_1)$, $A_1 \sin(\omega_n t + \varphi_1)$,

$(A_1 + A_2t)e^{\lambda_1 t}, A_1 e^{\lambda_1 t} + A_2 e^{\lambda_2 t}$, (c) third order responses that are considered are only of $A_1 e^{\lambda_1 t} \sin(\omega_1 t + \varphi_1) + A_2 e^{\lambda_2 t}$ type and (iv) fourth order responses that are considered limited to $A_1 e^{\lambda_1 t} \sin(\omega_1 t + \varphi_1) + A_2 e^{\lambda_2 t} \sin(\omega_2 t + \varphi_2)$ type.

Also, it is noted that, for reasons of explainability, the NN's used herein are kept shallow. It has been found that in deep learning, both empirically and theoretically, the explainability is low [17]. In the same piece of literature, various definitions of explainability are presented; herein, the link to providing responses to the "how" question is adopted.

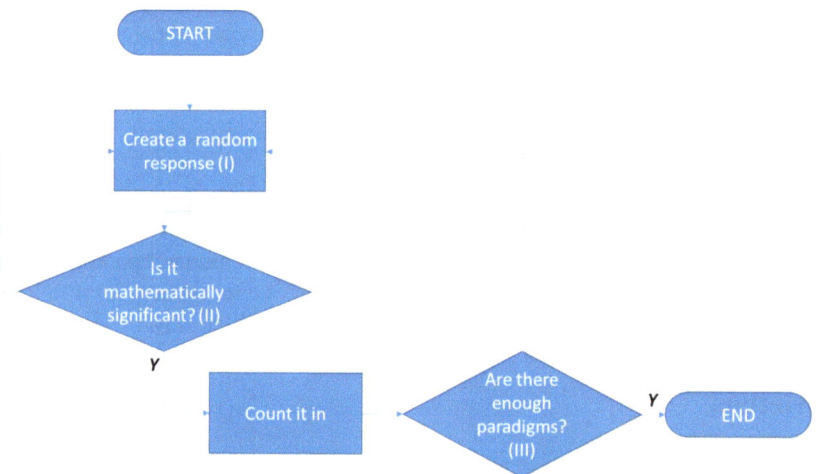

Fig. 1. Method of generation of datasets.

3   Results and Discussion

This section summarizes the results of the estimators.

Firstly, the case of Naïve Perceptron is tested, of the form *Output* $=$ $A[\,signal\ signal^2\,]$. Such an approach is based on output to classify a signal [18]. In this case, an augmentation in space has been made, adding the squared signal. This is an addition that was found empirically, through trial and error. The case of 5000 first-order responses and 5000 second-order response is tested. As shown in Fig. 2 (left), it is evident that even though the Perceptron works well for the case of first order, there are some misclassified responses of second order. Then, a traditional AIC estimator is used (Fig. 2, right). The AIC estimation is very fundamental and is based on taking into account the likelihood from stochastic estimation theory [19] but also includes some type of penalty for an increasing estimated order. It seems it works better than a Perceptron estimator, however, there are still some cases of misclassification. In Fig. 2, the responses of the two aforementioned classifiers are shown. The response is for the two classes of first and second order responses are $(1, -1)$ for the Perceptron and $(1, 2)$ for the AIC estimator.

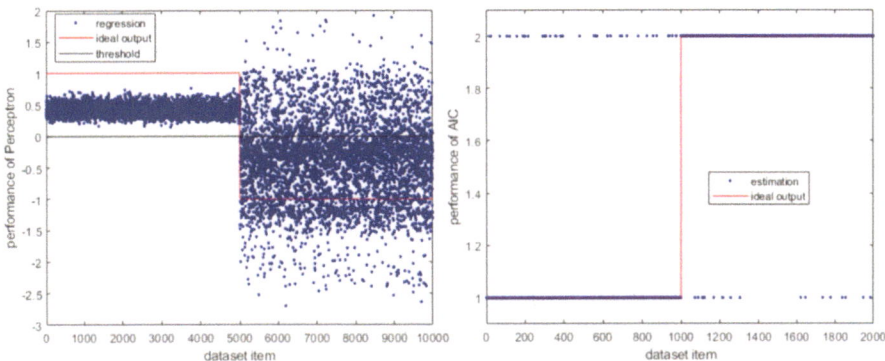

Fig. 2. Naive Perceptron results on identification (left) and AIC estimation outcome (right).

In the latter case, the order itself is the output, after the AIC metric has been estimated for both classes.

It is worth noting that a NN of a single hidden layer of 5 neurons has the best performance (Fig. 3). As a matter of fact, 0.74 s were required for the AIC estimator, while the NN required only 0.012 s (for a single decision). In any case, such a shallow NN gives the opportunity for explainability and is highly fast in its response. Figure 3 summarizes the NN output (1, −1), for first and second order responses, respectively.

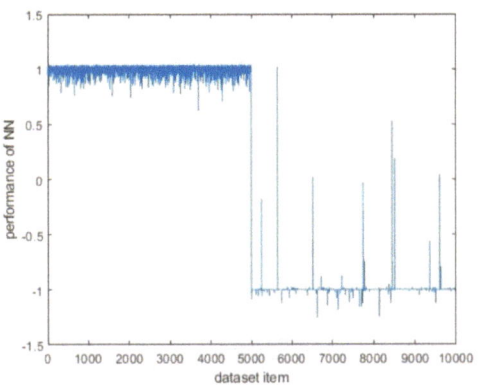

Fig. 3. NN results for order estimation.

It seems that the reason the Perceptron classifier failed is the numerical sensitivity. So, to this end, the NN was checked against two classes, in two cases; in the first one, the numerical differences between the classes (i.e. the constants in the responses) was large, while in the second one it was smaller. For different configurations of the NN, namely the number neurons per hidden layer, which was as per below: [5, 10, 15, 20], the regression R metric is equal to (0.945, 0.984, 0.993, 0.98) and (0.977, 0. 9957, 0.9957, 0.9957), respectively. However, here, the limit of deep learning has been reached; in any case to avoid any fallacies, a statistical analysis is required.

3.1 Capabilities of the NN's in the Case of Multiple Classes

As a natural continuation of Sects. 3.1 and 3.2, the performance of NN's in the case of multiple classes was checked. The superimposed poles' placement diagrams express the complexity/geometry of the responses and are used as resemblance indicators by people who are familiar with them; this indicates potential knowledge transfer, and hence explainability. It is noted, however, that the modelling effort is not integrated in such a metric, despite the fact that AIC takes it into account.

Regarding the training, the maximum fails (validation points) have been considered to be 120, while the minimum gradient for training has been 1e−10. The configuration [6, 12] was used, as the geometry of the responses is related not only to the decay but also to curvature. The choice could be characterized as heuristic, however, a brief investigation can be found in the appendix. Also, the low number of coefficients is crucial for the training time. The results are shown in Fig. 4; it appears that the complexity of the geometry of second order responses is quite high, due to curvature. In general, the main limitations of the current methodology are two. The first one is that the explainability requirement prohibits the NN's to reveal their true capabilities for performance. Also, Fig. 4 is only indicative of the NN's and statistical metrics should be presented, as the repetitive training of the NN's, due to the stochastic character of training can be considered part of the modelling effort. However, the step of showing the potential of NN's in order identification, even for small orders, has been achieved.

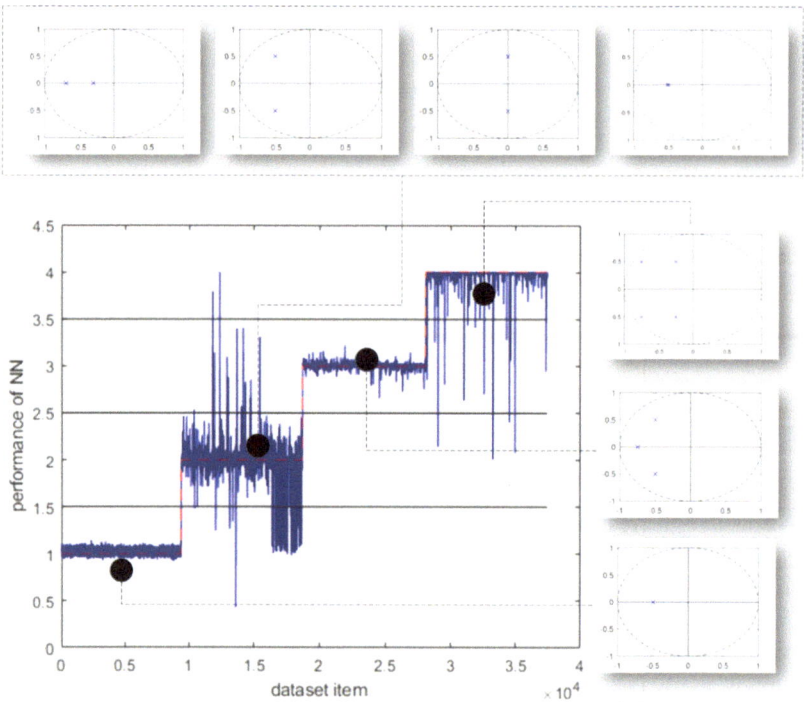

Fig. 4. A typical response of a NN with four classes, based on the number of poles.

4   Conclusions and Future Outlook

Coming to conclusions, the NN's are promising in terms of automated identification of ARX equations. However, the knowledge embedded in them seems to have limitations, due to numerical differences taken into consideration. Intuition can be built from the conducted experiments, i.e. hierarchical classification can be even more effective; grouping of the cases and then more focused classification. However, more human-in-the-loop identification may be required, so that patterns are extracted, and investigating the relation to sampling. This way, there will be mutual knowledge transfer between human and AI. Also, the following steps should be considered: working out numerical differences so that per configuration, there is complete separability, elaborating data and knowledge libraries, so that practices for shaping NN's are able to be generated (other AI techniques could also be considered), explainable deep learning being taken into consideration towards such matters and Finally, as an extra step, working with images, instead of signals, either for 2D systems or in terms of time-frequency representations.

Acknowledgement. The current work has been supported by EU project BRIDGES 5.0 (101069651).

Appendix: Capabilities of the NN's in the Context of Geometry

This section involves checking the performance of NN's with respect to their meta-parameters. In Fig. 5 (top), two arbitrary classes were tested (red vs. green); their threshold has been a sinusoidal curve to emulate the responses curvature. The results rather conclusive: the complexity was addressed adequately by the configuration [6, 12]. Figure 5, in general, involves two axes: error vs configuration, as represented by the NN structure and the classes geometry. For the case in the top picture that is discussed here, the data size considered is ten times the number of NN coefficients, while 5000 epochs, goal 1e−9, 10 + 0.3N checks and (1e−10)/N for gradient limit are regarded, with N being the size of dataset. The same study was repeated with more elaborated geometries, characterized by different topology (i.e. Betti numbers [20], indicating features such as cavities) (Fig. 5, bottom). The difference here is that the minimum dataset size considered here is 400, due to the geometry complexity. Results show a radical decrease from 0.1 to 0.01 utilizing different configurations, proving that the dimensionality of the data (e.g. 2 here) does not play a significant role here. It is also verified that the high number of neurons can be more representative of complicated geometries [21].

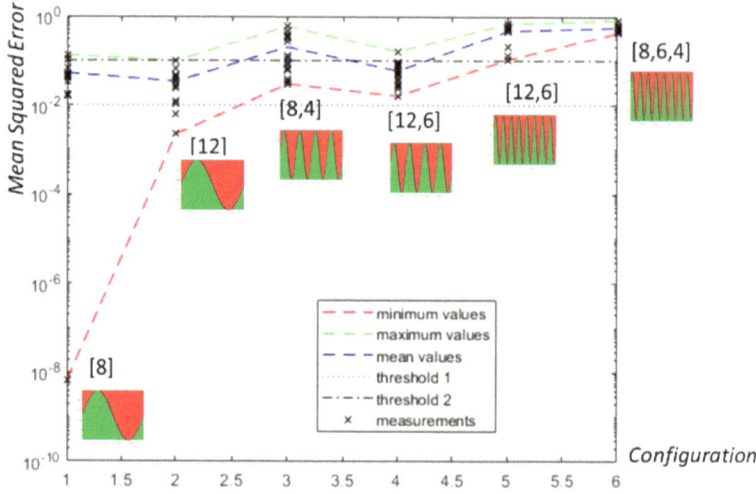

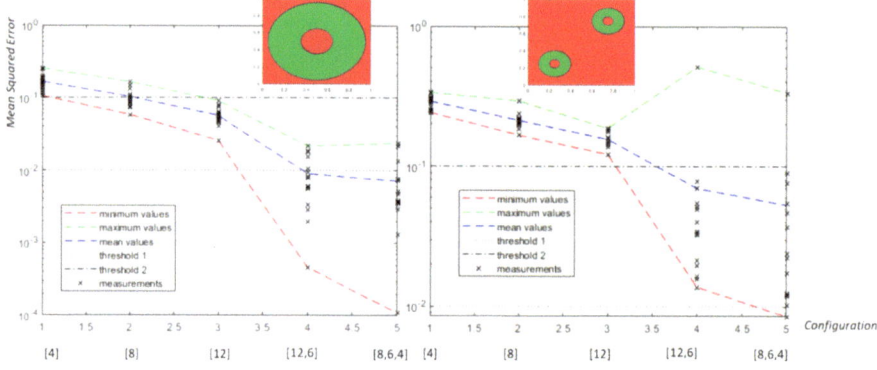

Fig. 5. Mean square error for different classes and different NN configurations: Sinusoidal classes (top) and complicated classes with cavities (bottom).

References

1. Stavropoulos, P., et al.: AI in manufacturing and the role of humans: processes, robots, and systems. In: Handbook of AI at Work, pp. 119–141. Edward Elgar Publishing (2024)
2. Rossi, A., Moretti, M., Senin, N.: Neural networks and NARXs to replicate extrusion simulation in digital twins for fused filament fabrication. J. Manuf. Proc. **84**, 64–76 (2022)
3. Stavropoulos, P., Papacharalampopoulos, A., Michail, C.K., Chryssolouris, G.: Robust additive manufacturing performance through a control oriented digital twin. Metals **11**(5), 708 (2021)
4. Wei, Z., et al.: Reliability optimization of the honeycomb sandwich structure based on a neural network surrogate model. Materials **16**(23), 7465 (2023)
5. Avendaño-Valencia, L.D., Chatzi, E.N.. Spiridonakos, M.D.: Surrogate modeling of nonstationary systems with uncertain properties. In: Proceedings of the European Safety and Reliability Conference ESREL (2015)

6. Vijayaraghavan, S., et al.: A data-driven reduced-order surrogate model for entire elastoplastic simulations applied to representative volume elements. Sci. Rep. **13**(1), 12781 (2023)
7. Li, S., et al.: An adaptive SVD–Krylov reduced order model for surrogate based structural shape optimization through isogeometric boundary element method. Comput. Methods Appl. Mech. Eng. **349**, 312–338 (2019)
8. Papacharalampopoulos, A.: Investigating data-driven systems as digital twins: numerical behavior of Ho–Kalman method for order estimation. Processes **8**(4), 431 (2020)
9. Kritzinger, W., et al.: Digital Twin in manufacturing: a categorical literature review and classification. Ifac-PapersOnline **51**(11), 1016–1022 (2018)
10. Barriga, R., Romero, M., et al.: Advanced data modeling for industrial drying machine energy optimization. J. Supercomput. **78**(15), 16820–16840 (2022)
11. Rewieński, M., Fotyga, G., Lamecki, A., Mrozowski, M.: Automated reduced model order selection. IEEE Antennas Wirel. Propag. Lett. **14**, 382–385 (2014)
12. Alqawasmi, K.E., Alsmadi, A.M.: Estimation of ARMA model order using artificial neural networks. Circuits Syst. Signal Process. **42**(7), 4129–4147 (2023)
13. Ottoni, A.L.C., Souza, A.M., Novo, M.S.: Automated hyperparameter tuning for crack image classification with deep learning. Soft. Comput. **27**(23), 18383–18402 (2023)
14. Stavropoulos, P., Papacharalampopoulos, A., Souflas, T.: Indirect online tool wear monitoring and model-based identification of process-related signal. Adv. Mech. Eng. **12**(5), 1687814020919209 (2020)
15. Papacharalampopoulos, A., Christopoulos, D., Stavropoulos, P.: Towards a surrogate spatiotemporal model of additive manufacturing for digital twin-based process control. Procedia CIRP **121**, 73–78 (2024)
16. Proakis, J.G., Manolakis, D.G.: Digital Signal Processing: Principles, Algorithms, 3rd ed. Prentice Hall (1995)
17. Herm, L.V., Heinrich, K., Wanner, J., Janiesch, C.: Stop ordering machine learning algorithms by their explainability! A user-centered investigation of performance and explainability. Int. J. Inf. Manage. **69**, 102538 (2023)
18. Serin, G., Sener, B., et al.: Deep multi-layer perceptron based prediction of energy efficiency and surface quality for milling in the era of sustainability and big data. Procedia Manuf. **51**, 1166–1177 (2020)
19. Urniezius, R., Kemesis, B., Simutis, R.: Bridging offline functional model carrying aging-specific growth rate information and recombinant protein expression: entropic extension of akaike information criterion. Entropy **23**(8), 1057 (2021)
20. Giri, S.K., Mellema, G.: Measuring the topology of reionization with Betti numbers. Mon. Not. R. Astron. Soc. **505**(2), 1863–1877 (2021)
21. Alfarra, M., Bibi, A., et al.: On the decision boundaries of neural networks: a tropical geometry perspective. IEEE Trans. Pattern Anal. Mach. Intell. **45**(4), 5027–5037 (2022)

AI in Equipment Level

Leveraging Generative AI for Synthetic Data Generation: Improving 6-DOF Pose Estimation in Assembly Systems

Christos Konstantinou, Nikos Kampouroglou, Nikos Theodoris, Fotis Basamakis, Christos Gkournelos, and Sotiris Makris[✉]

Laboratory for Manufacturing Systems and Automation, University of Patras, Rion Patras 26504, Greece
makris@lms.mech.upatras.gr

Abstract. In recent years, accurate 6-DOF (six degrees of freedom) pose estimation has emerged as a pivotal technology in manufacturing, enabling the precise localization and manipulation of objects in complex environments. The effectiveness of 6-DOF pose estimation algorithms critically depends on the availability of diverse, well-annotated datasets. However, obtaining and annotating such datasets present significant challenges due to their scarcity and the intensive labor required for accurate labeling. To address these issues, we propose an innovative approach that employs synthetic data generation, powered by generative artificial intelligence (AI) techniques specifically tailored for industrial applications. Our method enhances the synthetic data generation process by utilizing generative adversarial networks (GANs), which infuse the data with contextual details relevant to manufacturing environments. This process is further augmented by advanced rendering techniques and simulations that create realistic industrial scenes, complete with accurately annotated ground truth for 6-DOF poses. We validate the effectiveness and robustness of our proposed solution through its application in a real-world industrial use case, demonstrating its potential to substantially improve 6-DOF pose estimation in a manufacturing case, used for robotic picking of electronic parts.

Keywords: Synthetic Dataset Generation · Generative Adversarial Networks · Machine Learning · 6D Pose Estimation

1 Introduction

Manufacturing environments present unique challenges [4] to pose estimation algorithms, particularly in cluttered scenes characterized by disorganized backgrounds, occlusions between objects, and changes in lighting conditions. The development of robust 6-DOF pose estimation models comes with significant challenges, primarily due to the reliance on extensive, accurately annotated datasets [8]. Within manufacturing environments, products and industrial parts

© The Author(s) 2025
K. Alexopoulos et al. (Eds.): ESAIM 2024, LNME, pp. 71–79, 2025.
https://doi.org/10.1007/978-3-031-86489-6_8

diverge from the commonplace objects typically found in large open datasets. Industrial objects often exhibit unique characteristics like complex geometries and uncommon material textures. Thus, they cannot be used seamlessly in manufacturing applications, creating the need for a systematic framework for data generation.

As highlighted in recent studies, expanding existing datasets with synthetic data, has proved to be a promising strategy to overcome the limitations raised by the absence of physical industrial data. Based on latest research in large models, Generative adversarial networks (GANs) can be a promising approach to overcome such constrains, providing a virtually limitless pool of annotated data by adding contextual details relevant to manufacturing environments. GANs can be used to create synthetic datasets that are tailored to replicate complex, real-world scenarios with remarkable accuracy, leading to pose estimation models that are both flexible and robust after appropriate training[11]. However, the transition from synthetic to real-world application presents its own set of challenges. The "reality gap" [12], a term denoting the discrepancy between model performance in synthetic versus real environments-remains a significant obstacle [2].

Recent literature underscores a range of approaches, each addressing unique challenges within the domain as it can be seen summarized in Table 1. In general, estimating 6D poses from RGB images presents a number of challenges. Perspective ambiguities, wherein objects exhibit similar appearances from varying viewpoints, hinder effective learning, especially in cluttered scenarios [2][3]. Moreover, environmental factors such as lighting variations and complex backgrounds further complicate the algorithmic performance.

Table 1. Comparison of Various 6D Pose Estimation Approaches

Reference	Real-Time	Uses RGB	Uses Depth	Cluttered	Refinement	Real DT	Synth. DT	Implementation	Testing Dataset
PoseCNN [14]		✓		✓		✓	✓	TensorFlow	YCB-V, LineMOD
DOPE [12]	✓	✓			✓			PyTorch	YCB-V
G2L-Net [3]	✓	✓	✓		✓			PyTorch	YCB-V, LineMOD
Megapose6D [6]		✓	✓	✓	✓		✓	PyTorch	BOP datasets , ModelNet
SAM6D [7]		✓	✓	✓	✓		✓	PyTorch	BOP datasets
FoundationPose [13]	✓	✓	✓	✓	✓		✓	PyTorch	BOP datasets , YCBInEOAT

Despite the innovative characteristics of all these mentioned approaches, notable gaps persist within the current landscape of 6D pose estimation approaches. The integration of synthetic datasets, while beneficial for training due to many factors [1], such as time efficiency in data generation and collection, they introduce challenges related to domain adaptation and real-world generalization. Furthermore, achieving real-time performance without compromising accuracy and robustness, in terms of pose estimation, remains an ongoing challenge, particularly in cluttered manufacturing environments.

In summary, there exists a need for further research to address the inherent challenges and bridge the gap between synthetic training environments and real-world application scenarios. In context, this paper proposes a framework that can identify 6-DOF poses of novel objects, based solely on their Computer Aided Design (CAD) files, and in a textual description of their external visual characteristics. For further enhancing the detection precision, an automated way of generating the bounding boxes of these novel objects was implemented based upon a synthetic generated dataset. This research paper is organized as follows: In Sect. 2 the overall approach structure is defined, which is subsequently addressed in Sect. 3 where the implementation details are presented. The application of the proposed synthetic data generation for 6D pose estimation is evaluated in the Sect. 4 using a real industrial use case. Finally, in Sect. 5 the conclusions and future work are reported.

2   Approach

An approach combining GANs, CAD models and advanced simulation techniques has been developed in an attempt to construct synthetic datasets dedicated to 6-DOF pose estimation in manufacturing environments.This approach created detailed and varied training data that enhances the model's ability to accurately estimate poses of industrial-oriented objects.

As depicted in Fig. 1, the initial stage of the process involves the utilization of a pretrained GAN model [11] dedicated to the background image generation.

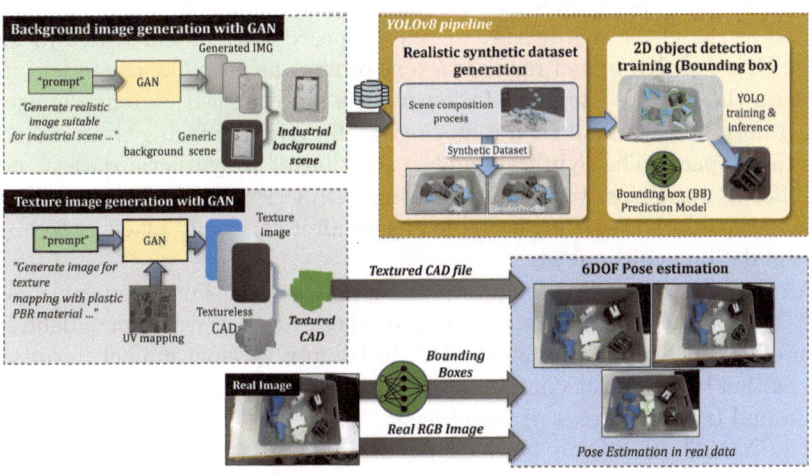

Fig. 1. Overview of the proposed framework

Since these type of models are trained on a vast amount of diverse environments, they can generate background images such as industrial floors, assembly lines and warehouses. This network uses both text and image prompts to accurately render the requested environment and its intricate patterns. The text prompt specified the desired outcome, such as "Produce high-fidelity background images of industrial rug featuring light-blue hues and multiple circular patterns."

Following the background generation process, another instance of a GAN was set up, aimed specifically for object texture generation. This model uses the object's exported UV mapping as a reference image, and a textual description outlining the desired texture characteristics. Both inputs fed into the network for it to generate an accurate texture that can wrap the object's geometry. This technique achieves a high degree of visual fidelity within the synthetic scenes. In this GAN, a text prompt guides the network to generate textures that mimic the desired material industrial characteristics. Using prompts such as "Generate high-quality image suitable for texture mapping focusing on plastic PBR material", it was possible to generate several accurate texture images, as can be seen in Fig. 1, leading to the synthetic dataset containing the objects of interest.

The integration of background images and detailed CAD models, in combination with advanced simulation techniques such as physics-based modeling, varying lighting conditions, and dynamic camera movements, generates synthetic scenes annotated with 6-DOF poses. These scenes exhibit a high degree of variety and realism, closely mirroring actual manufacturing environments. To automate this process and minimize human intervention, a YOLO based object detection system was developed and trained on the generated synthetic dataset. It successfully produced 2D bounding boxes for the physical components, which were subsequently used as fine-tuning inputs for the pose estimation model.

3   Implementation

In order to add photorealism to our data, Stable Diffusion XL model [9] was selected as the GAN architecture for the proposed work as illustrated in Fig. 2. Stable Diffusion XL model requires a text prompt as an input to render images based on the provided text. Diffusion models are designed to refine images by adding noise to them and then removing it, effectively diffusing the noise across the image space. The synthesis and generation of a synthetic dataset covering different scenarios, was achieved using BlenderProc2 [5]. This is a Blender pipeline capable of rendering realistic images after randomly placing the objects in a simulation environment.

All the physics-based calculations are applied to the simulation environment via a Python API. In this study, the synthetic data generation pipeline was executed on an Nvidia RTX 3060 GPU, in Ubuntu 22 environment, resulting in the creation of a dataset consisting of 50,000 synthetic images accompanied by the ground truth 6D poses of the objects.

The YOLO (v8) (You Only Look Once) object detection framework, as described in [10], was trained on the generated synthetic dataset with the 50,000

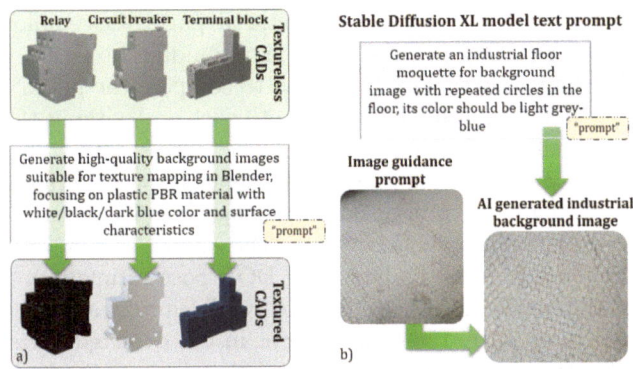

Fig. 2. a)Text-to-image texture generation using GANs b) Background image generation using GANs

images to provide the 2D bounding boxes of the industrial objects given an unseen image. The training procedure included 100 epochs on the synthetic dataset with a ratio of 70-15-15% for training, validation and test images respectively. The extracted bounding box was essential for the subsequent stages of the pose estimation process, thus enhancing the accuracy and automation of the system.

Regarding the 6D pose estimation, the MegaPose model [6], was utilized for estimating the translation and orientation of the objects. This model plays a benchmark role in order to demonstrate the reliability of the proposed method. As depicted in Fig. 1, the system receives as an input a real (not previously known) image of the region of interest, a CAD file with no specific texture of the part of interest, and two textual descriptions. One for the external visual characteristics of the industrial part, and the second for the surrounding environment. As it will be described in the following sections the evaluation of this method was performed both in the synthetic and the real domain.

4 Case Study

The presented work for the generation of an estimated 6D pose, has been deployed and tested into two use cases, that involve the detection and the handling of electrical parts placed randomly in a bin. These parts include terminal blocks, relays and circuit breakers as can be seen in Fig. 3. The proposed application for incorporating synthetic datasets for 6-DOF pose estimation is crucial for manufacturing, as it significantly augments the accuracy of object localization and manipulation in complex industrial environments. This field faces significant challenges due to the occlusions, varying lighting conditions, and diverse geometries of industrial parts, which complicate accurate pose estimation. The first case study involves comparing the impact of the GAN-textured CAD objects to texture-less CAD on the results of pose estimation algorithm.

Furthermore, the synthetic dataset of electrical parts was employed to train the YOLO model, enabling it to generate 2D bounding boxes for the detected parts. These bounding boxes were then used to compare the performance of the Megapose6D method on synthetic images, both with and without the fine-tuning provided by the bounding box input.

Following the proposed approach, a simulation was performed through Blenderproc resulting in a synthetic dataset comprising 50 scenes and each scene containing 100 frames. In order to evaluate the impact of GAN-generated textures on the produced synthetic data for 6-DOF pose estimation, metrics such as the angular difference of quaternions and Euclidean distance between translations are used to quantify the enhancement in the detected parts' poses estimation. Results were categorized based on whether the synthetic data included GAN-generated textures or not, and whether real data was used with or without GAN textures. The findings revealed a notable improvement in performance with the application of GAN textures. When the CAD models were overlayed by the GAN generated textures, a noticeable increase in the pose estimation accuracy was observed. On the other hand, the accuracy remained almost the same. Moreover, the impact of GAN textures was even more noticeable when analyzing real-world data. These results, as it can be observed at Table 2, demonstrate that GAN-generated textures enhance the accuracy and robustness of pose estimation models, especially in scenarios involving complex and varied textures.

Table 2. Comparison of pose estimation accuracy with and without GAN textures

	With GAN texture	Without GAN texture	Difference
Synthetic Data			
Blue terminal block	85.92%	85.90%	+0.02%
White circuit breaker	83.06%	78.12%	+4.94%
Black relay	81.12%	80.07%	+1.05%
Real Data			
Blue terminal block	92.43%	74.50%	+17.93%
White circuit breaker	95.55%	74.30%	+21.25%
Black relay	80.44%	80.43%	+0.01%

Similar pose estimation tests were carried out with real images of electrical parts. The results of the pose estimation followed the previous logic, with lower pose scores presented in occluded conditions or when the bounding box was not sufficiently precise to indicate the exact boundaries of the object. These scores were lower than those for synthetic objects, ranging from 65% to 80%. Realistic object textures from GAN were added to the synthetic data, along with an extracted bounding box from the YOLO trained model. The predicted bounding box of the electric part, as a result from the YOLO training on the dataset, achieved markedly higher pose scores, reaching 95% to 98%, even when the object was not easily disguised from the environment, as illustrated in the Fig. 3.

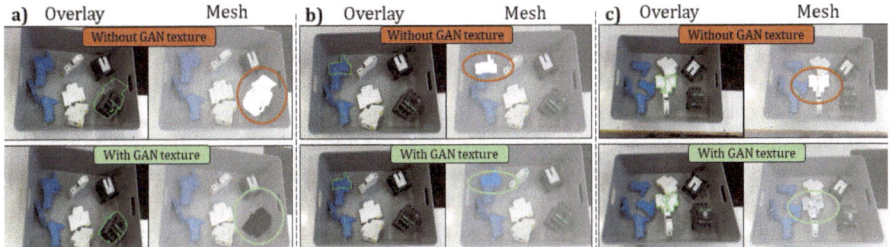

Fig. 3. Pose estimation results on real sample

5   Conclusion

In this paper, a synthetic dataset enhanced by GAN-generated textures is presented to improve pose estimation. The distinguish feature compared with novel 6-DOF pose estimation models, is the utilization of GAN generated content that augmented key characteristics of the approach, mimicking real industrial parts, and environments. These characteristics consist of the textures of the CAD models, and the industrial background spaces. For further enhancing the estimation results, a YOLO based object detection network was implemented, trained upon the GAN generated images and scenes. The proposed method was tested and evaluated in a real industrial use case, consisting of a set of 3 different components: a term block, a circuit breaker and a relay switch. Using the CAD files of the parts, and real images taken from a top-view camera, this approach was able to correctly identify the poses of all the components and even successfully handling the random occlusions. Comparisons with and without assist from the GAN image generation were performed, demonstrating significant improvements in the pose estimation accuracy, indicating an almost 18% increase. In conclusion, this framework highlights the effectiveness of GAN content generation, and proves its usability in industrial environments and complex use cases. However, despite achieving high evaluation scores, a number of challenges arouse. Future work aims to create an ecosystem of tools that finalize the application of an autonomous bin-picking, achieving an end-end interaction from the robotic environment. Finally, smart 3D scanner technologies, reconstructing the CAD models of industrial parts, can be utilized, further simplifying the overall flow.

Acknowledgments. This research has been supported by the EC funded projects "MASTERLY: Nimble Artificial Intelligence driven robotic solutions for efficient and self-determined handling and assembly operations" (Grant Agreement: 101091800) and "RENÉE: Flexible remanufacturing using AI and advanced Robotics for circular value chains in EU industry (Grant Agreement: 101138415)"

References

1. Abufadda, M., Mansour, K.: A survey of synthetic data generation for machine learning. In: 2021 22nd International Arab Conference on Information Technology (ACIT), pp. 1–7. IEEE (2021)
2. Cao, H., Dirnberger, L., Bernardini, D., Piazza, C., Caccamo, M.: 6IMPOSE: bridging the reality gap in 6d pose estimation for robotic grasping (Mar 2023). http://arxiv.org/abs/2208.14288, arXiv:2208.14288
3. Chen, W., Jia, X., Chang, H.J., Duan, J., Leonardis, A.: G2L-Net: Global to Local Network for Real-time 6D Pose Estimation with Embedding Vector Features (2020). https://arxiv.org/abs/2003.11089
4. Chryssolouris, G.: Manufacturing systems: theory and practice. Mechanical engineering series, Springer, New York, 2nd ed edn. (2006), oCLC: ocm61253973
5. Denninger, M., Winkelbauer, D., Sundermeyer, M., Strobl, K.H., Humt, M., Triebel, R.: BlenderProc2: A procedural pipeline for photorealisticrendering. J. Open Source Softw. **8**(82), 4901 (2023). https://doi.org/10.21105/joss.04901
6. Labbé, Y., et al.: MegaPose: 6D Pose Estimation of Novel Objects via Render & Compare (Dec 2022). http://arxiv.org/abs/2212.06870, arXiv:2212.06870 [cs]
7. Lin, J., Liu, L., Lu, D., Jia, K.: SAM-6D: Segment Anything Model Meets Zero-Shot 6D Object Pose Estimation (Mar 2024). http://arxiv.org/abs/2311.15707, arXiv:2311.15707 [cs]
8. Makris, S.: Cooperating Robots for Flexible Manufacturing. Springer International Publishing, Cham (2021)
9. Peebles, W., Xie, S.: Scalable Diffusion Models with Transformers (Mar 2023). arXiv:2212.09748 [cs]
10. Redmon, J., Divvala, S., Girshick, R., Farhadi, A.: You Only Look Once: Unified, Real-Time Object Detection (May 2016). arXiv:1506.02640 [cs]
11. Rojtberg, P., Pollabauer, T., Kuijper, A.: Style-transfer GANs for bridging the domain gap in synthetic pose estimator training. In: 2020 IEEE International Conference on AIVR, pp. 188–195. IEEE, Utrecht, Netherlands (Dec 2020). https://doi.org/10.1109/AIVR50618.2020.00039
12. Tremblay, J., To, T., Sundaralingam, B., Xiang, Y., Fox, D., Birchfield, S.: Deep Object Pose Estimation for Semantic Robotic Grasping of Household Objects (Sep 2018). arXiv:1809.10790 [cs]
13. Wen, B., Yang, W., Kautz, J., Birchfield, S.: FoundationPose: Unified 6D Pose Estimation and Tracking of Novel Objects (2023). https://doi.org/10.48550/ARXIV.2312.08344, https://arxiv.org/abs/2312.08344
14. Xiang, Y., Schmidt, T., Narayanan, V., Fox, D.: PoseCNN: a convolutional neural network for 6D object pose estimation in cluttered scenes (May 2018). arXiv:1711.00199 [cs]

CPPM Copilot: Proposing an AI-Based Assistant for Manual Assembly Tasks in a Flexible Production

Jonathan Nussbaum[1]([✉]), Tatjana Legler[1,2], and Martin Ruskowski[1,2]

[1] Chair of Machine Tools and Control Systems, University of Kaiserslautern-Landau (RPTU), 67663 Kaiserslautern, Germany
jonathan.nussbaum@rptu.de

[2] Innovative Factory Systems, German Research Center for Artificial Intelligence (DFKI), 67663 Kaiserslautern, Germany

Abstract. With an increase in flexibility in industries adopting the tenets of shared, skill-based, and modular production, a higher variability of products a single manufacturer offers can be anticipated. An immediate consequence is a potential inability to instruct assembly personnel in an adequate and detailed manner. This problem would primarily affect small and medium enterprises engaged in individual and small series productions. To face the issue, this paper examines the suitability of facilitating an AI assistant to support workers handling an ever-expanding range of assembly tasks. This assistant would be realized through a retrieval-augmented generation system, founded on an Large Language Model (LLM) and a knowledge base. We especially propose a locally trained and hosted LLM, aiming to enhance effective flexibility and applicability while minimizing individual installation and setup times. By strictly formalizing descriptions of assembly or disassembly steps in a knowledge base, manufacturing difficulties can be presented as informational problems, at which LLMs excel. Through this, we are trying to extend the increased efficiency of knowledge workers empowered by utilization of LLMs such as GPT-4 into manufacturing. Using verbal inputs as well as reading the generated feedback back to a worker, we aim to keep a worker engaged with their primary tasks and, furthermore, reduce idle times caused by knowledge gaps. By taking this verbal/auditive-only approach, we secondarily aim towards increasing worker autonomy by answering miscellaneous workplace-related questions alongside knowledge-based problems of the day-to-day business.

Keywords: Worker assistance · Large Language Models · AI in manufacturing · Capability-Skill-Service model

1   Introduction and Motivation

Humans still hold one of the most important positions inside modern manufacturing shop floor environments, mainly due to their adaptability, their versatility as well as their problem-solving skills. While appearances like *Figure's*

© The Author(s) 2025
K. Alexopoulos et al. (Eds.): ESAIM 2024, LNME, pp. 80–88, 2025.
https://doi.org/10.1007/978-3-031-86489-6_9

Figure01 [15] or the new iteration of *Boston Dynamics'* Atlas [4] are aiming to replicate the general dexterity of humans, neither robot is supposed to replace a human in the complex and flexible sectors that are small series or even individual part production, with Figure aiming to support the warehouse and retail workforce and Boston Dynamics being factored in for automotive serial production [5,7]. Neither these two humanoid robots, nor conventional stationary robots, can replace the skilled personnel required to operate agile production lines [8]. Still, these skilled workers face problems with the ever-changing product palette of the modern, flexible shop floors.

Within this paper, we are concerned with just the small-scale problem that is manual product assembly, within which we are focusing on issues regarding assembly instructions. Key challenges regarding manual assembly have been identified by [9] as

- instructions being of poor quality and too general in their nature,
- a detailed, digital documentation being unavailable to staff facing problems,
- and conventional paper-based instructions lacking traceability.

This is by no means an argument for extraordinarily formalized assembly instructions, as a recent study has found human workers preferring to pick a cognitively demanding task, while allocating manual labor to a cobot [12].

Regardless of whether there is a cobot to take up other, repetitive tasks, the mentally demanding tasks of figuring out new and not yet learned assembly instructions is an adequate example for a relevant problem-solving skill. But while a human might more or less enjoy the time spent working out how to assemble a new product, they might benefit from an assistant supporting them. While not solving the workers' problems for them, a smart personal assistant can still be of benefit by increasing problem-solving skills in the long term, as shown by [14]. If applicable, this would enable workers to more easily comprehend instructions initially found to be rather incomprehensible. On the other hand, if there is a possibility to formalize and standardize assembly instructions to a point, where the manual assembly itself is solely considerable to be a problem of knowing the next steps, then the findings of [2] regarding the increase in efficiency of knowledge workers utilizing Generative Artificial Intelligence (GenAI) can be transferred into manual assembly.

2   Context and Background

Following, we highlight the background, concepts, and technologies necessary for our concept. Next to the benefits of verbally communicating assistants, we cover the Capability-Skill-Service model (CSS model) and GenAI as well as Natural Language Processing (NLP).

2.1   Benefits of a Verbally Communicating AI Assistant

In a study, [13] have found difficulties when integrating a worker assistance system relying on visual input for the personnel. Workers with cognitive disabilities sometimes forgot to use their assistance systems. Experienced workers even chose to disregard system instructions, while still appreciating the systems' potential for alerting them in cases of potential assembly errors. Furthermore, their study showed problems when using, for example, a Microsoft HoloLens or projectors for visual assistance, stating the HoloLens' limited battery life and a projectors' increased need for maintenance.

Considering an approach contrary to using visual interactions between personnel and an assistance system, a system could be interactive with on a voice-first basis. A voice-first approach would – in the case of worker assistance – mean using verbal outputs of a human as the primary system input. Advantages of such a voice controlled system are the flexibility to individually adapt them to any environment or worker, as well as being user-agnostic and therefore open to anybody willing to use it. [11]

Taking just these two positions into account and, furthermore, [11]'s statement of AI-based personal assistants having a high potential when a user would need to interface with machinery as well as [2]'s values regarding the increase in performance observed in knowledge workers when utilizing an AI assistant (12.2 % more tasks completed, being 25.1 % faster, while achieving 40 % better results), leads to the premise of an AI-based and voice controlled assistance system. Notably, the averaged values reported by [2] were drastically lower for already better-than-average workers, while still being a net positive, but – more importantly – the increase in performance was even higher for workers below the average initial performance levels.

2.2   Capability-Skill-Service Model

The CSS model is a conceptual framework that unifies the terminology and defines a vocabulary for capabilities, skills, and services in the context of production processes. A simplified overview of the model is shown in Fig. 1. It aims to enhance understanding and interoperability in new production concepts and support standardization activities in the manufacturing industry. The model is an extension of the Product-Process-Resource representation paradigm and focuses on capturing functions at different levels: Capabilities represent functions in production process steps; skills are implementations of these functions provided by specific resources; and services define offerings of capabilities in broader supply chain networks.

In the scope of the concept proposed in this work, the relevant parts of Fig. 1 are the offered **service**, which provides the offered **capability** and the **skill** realizing it, as well as the **resource** providing said capability and skill. From a top-down view, the service offered is an outside representation and offer towards other agents in need of the specific service. A capability, provided by the service and the resource, is a representation of an action, complete with options for parametrization and beforehand information regarding, example given, runtimes or, ever-more important, energy and material consumption or greenhouse gas emissions. As a capability is an abstract entity provided by said services and provided by certain resources but not tied to either definitely, if there were multiple resources providing a certain capability, the service would be the provider

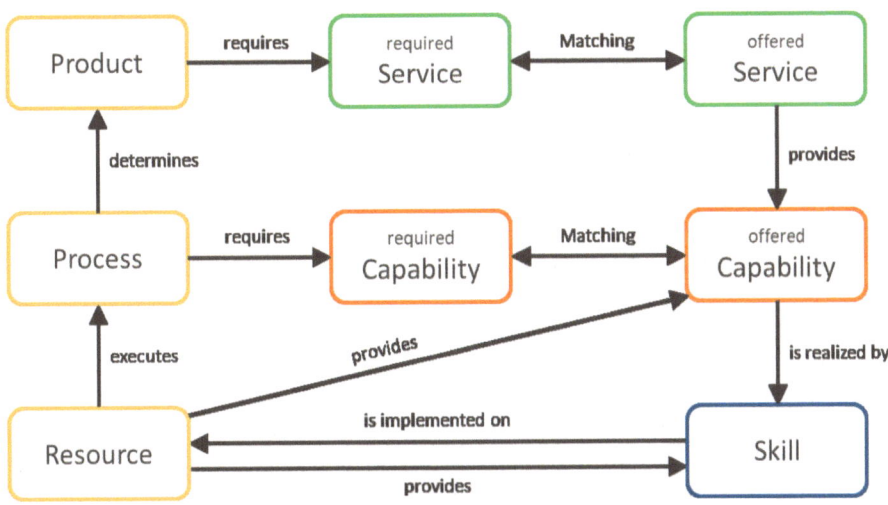

Fig. 1. Simplified overview on most important aspects of the CSS model [3].

for each of them. The implementation of a capability down into the machine and module level is the so-called skill [3].

2.3   Generative Artificial Intelligence and Natural Language Processing

Generative Artificial Intelligence is a term used colloquially for Artificial Intelligence (AI) capable of generating new information. Common tasks for GenAI stem from the field of NLP, taking the form of text interpretation and generation. Other media generated are songs, images, as well as videos. The generation follows a user input, which is often given in text form, optionally supplemented by documents, web pages, images or sound files. The topics covered by this diverge greatly, with a user being able to extract information about topics ranging from culinary advice, over an overview of Albrecht Dürer's Fechtbuch, through an explanation of general relativity fit for elementary school kids, to details regarding internationally less known tabletop role-playing games [6].

NLP is the field of automatic analysis and representation of human language. It is concerned with understanding, interpreting and generating human language on a meaningful scale. Further NLP tasks are, for example, speech recognition, translation, or summarization. [10]

Using AI networks for NLP tasks has been a long-going trend, with statistical language models and then neural language models being having been used for NLP tasks. Further development resulted in pre-trained language models trained on extensive datasets, the largest of which are known as **LLMs**. The emerging abilities of the models were: a sharp increase in general performance;

suddenly being able to generate very high-quality text samples; possessing robust learning; and reasoning abilities.

3   Proposed Concept

The generalized description of the concept is summarized in Fig. 2. A workers' spoken query is first picked up by a headset connected to the module and, using NLP, transformed into text. Semantic information is then extracted from this text by an LLM, before it is matched with information from the knowledge base, possibly utilizing knowledge graphs. From the knowledge base, information is retrieved and through the LLM expressed in natural text. This text is then turned into speech using NLP and played back to the worker. The separation of NLP and LLMs, as suggested in Fig. 2 is not strictly necessary, as LLMs tend to be generally efficient in NLP tasks. A separation might still prove useful, in case information modelling is a task utilized best by an LLM without additional tasks.

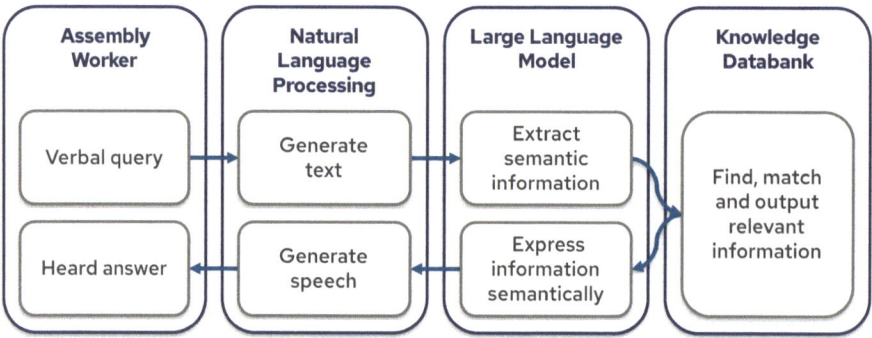

Fig. 2. Overview of the core concept.

3.1   Test Demonstrator and Demo Use-Case

During a previous project, 'KoKoBot – Setup of a collaborative and cooperative robotics platform', a demonstrator consisting of three Cyber-Physical Production Moduless (CPPMs) has been set up. The demonstrator and other project results can be seen in [1]. As one project requirement has been equipping the demonstrator with industrial-grade hardware, it will suit well as an assembly station, noting that one of the three modules is, in fact, called 'manual workstation'. An NVIDIA Jetson Orin already built into the demonstrator is meeting estimated computational requirements and will therefore be the hardware platform hosting the LLM and NLP applications. Additionally, in case the Jetson Orin proves to be a computational bottleneck, a workstation utilizing an RTX graphics card is part of the module as well, serving as an ample backup. Further equipment needed for the proof of concept would be a product to be used for the assembly or disassembly task. For this, the toy trucks used in the previous project would suffice. After a first proof of concept, more sophisticated assembly groups would undoubtedly be needed. Furthermore, any combination of microphone and speaker would suffice to capture vocal instructions and play back the assistant's replies.

3.2 Foreseeable Issues

When integrating an LLM a number of issues have to be addresses upfront. One of these is limiting the system's access to work-unrelated information. While the working personnel should be free to ask anything with a reasonable connection to their work, misconduct should preemptively be inhibited. Furthermore, hallucinations have to be addressed. For this, the proposed concept will initially be used as a test bed, closely monitoring the AI's behavior and enabling timely implementation of preventive measures.

3.3 Integration of the CSS Model

When integrating the CSS model with the above concept, the granularity of representing functions becomes a core question. One way of implementation would be to make the complete worker assistance into a single skill. In this case, an additional external input would cause the start of the event chain, beginning with the workers' verbal input. The skill would have no parametrization options, as well as not having any output except for the verbalized response of the system.a

Another, preferable, possibility would be to encapsulate most functionalities into singular skills, calling each other when in execution. This would enable the individual testing of each step of the general concept. Therefore, we propose splitting the functions into the following skills:

Skill 1: A skill that is always running and waiting for a command to start the next Skill.

Skill 2: The skill responsible for NLP in both directions, being configurable to either transcribe spoken text or verbalize textual input, either passing it on to skill three or playing it back to the human.

Skill 3: A skill taking text inputs and using them to fetch requested information from a database, while also capable of translating the database entries into human-readable text, which it transmits back to skill two.

Correct system behavior can thus be tested by using spoken as well as textual standard phrases with skill two, testing the correct textualization and verbalization, and calling the third skill either with a search string or a given database entry. The worker states a query, which is checked against the manual and a database, and results in a response by the module. Additionally, both skills can be used in other ways, with skill two also able to be used to give the user a heads-up in case of system-wide errors or warnings and skill three being available to different users for questioning the database.

Figure 3 shows a simplified version of this concept in action. Assuming skill one to be running and triggered by the command word, skill two is used to pick up the workers' query and call skill three. The relevant manual is identified and checked, as well as a database being consulted, before the information is sent back into skill two to be read out to the human.

Fig. 3. Simplified depiction of a worker stating a question, a manual and a database being consulted, and the system giving an answer.

4 Conclusion and Future Directions

Summarized, our concept is aiming to take an assembly workers' verbal input regarding a new product, the assembling steps of which are at least partially unknown. This verbal input is then compiled into a search query for a database containing information regarding the new part. The output of the database search is then compiled into natural language and read back to the human.

This endeavor is only a first step, with multiple possible next steps already figured out. Instead of taking the verbal input for a search query, it might also be a control sequence for the CPPM, triggering different functions and enabling more intuitive control of certain aspects of production. Moreover, the search queries don't necessarily have to relate to the part currently worked on, different other aspects of work might be included in the database, from a guide on how to correctly fill out applications for vacation time to information regarding the current menu in the canteen, stimulating a worker during otherwise possibly boring work and sating the need for certain information. Another aspect of the system is the relatability to different production environments. Human-Robot-Cooperation might be more easily coordinated, if the human can tell the robot what they will do next, as well as giving instructions on what the robot might be supposed to do. The system would likewise suit chemical or pharmaceutical laboratory environments processing a doctor's or pharmacists recipes as well as more classical shop floors, where part specifications could be checked on the fly without having to leave the working equipment to check the specification.

Acknowledgment. This work was funded by the Carl Zeiss Stiftung, Germany under the Sustainable Embedded AI project (P2021-02-009).

References

1. Project results - department of mechanical and process engineering at rptu (13052024). https://mv.rptu.de/en/fgs/wskl/forschung/kokobot/project-results
2. Dell'Acqua, F., et al.: Navigating the Jagged Technological Frontier: Field Experimental Evidence of the Effects of AI on Knowledge Worker Productivity and Quality (2023). https://doi.org/10.2139/ssrn.4573321
3. Diedrich, C., et al.: Information Model for Capabilities, Skills & Services: Definition of terminology and proposal for a technology-independent information model for capabilities and skills in flexible manufacturing. Plattform Industrie 4.0 (2022)
4. Dynamics, B.: Atlas — boston dynamics (4/17/2024). https://bostondynamics.com/atlas/
5. Dynamics, B.: An electric new era for atlas — boston dynamics (4/17/2024). https://bostondynamics.com/blog/electric-new-era-for-atlas/
6. Epstein, Z., et al.: Art and the science of generative AI. https://doi.org/10.1126/science.adh4451
7. FigureAI: About us — figure (4/19/2024). https://www.figure.ai/about-us
8. Freire, S.K., et al.: A cognitive assistant for operators: Ai-powered knowledge sharing on complex systems. IEEE Pervasive Comput. **22**(1), 50–58 (2023). https://doi.org/10.1109/MPRV.2022.3218600
9. Johansson, P.E.C., et al.: Challenges of handling assembly information in global manufacturing companies. J. Manuf. Technol. Manag. **31**(5), 955–976 (2020). https://doi.org/10.1108/JMTM-05-2018-0137
10. K., M., et al.: A survey (nlp) natural language processing and transactions on (nnl) neural networks and learning systems. E3S Web Conf. **430**, 01148 (2023). https://doi.org/10.1051/e3sconf/202343001148
11. Mark, B.G., Rauch, E., Matt, D.T.: Worker assistance systems in manufacturing: a review of the state of the art and future directions. J. Manuf. Syst. **59**, 228–250 (2021). https://doi.org/10.1016/j.jmsy.2021.02.017
12. Schmidbauer, C., et al.: An empirical study on workers' preferences in human-robot task assignment in industrial assembly systems. IEEE Trans. Human-Mach. Syst. **53**(2), 293–302 (2023). https://doi.org/10.1109/THMS.2022.3230667
13. Simões, B., et al.: Cross reality to enhance worker cognition in industrial assembly operations. Int. J. Adv. Manufact. Technol. **105**(9), 3965–3978 (2019). https://doi.org/10.1007/s00170-019-03939-0
14. Winkler, R., Söllner, M., Leimeister, J.M.: Enhancing problem-solving skills with smart personal assistant technology. Comput. Educ. **165**, 104148 (2021). https://doi.org/10.1016/j.compedu.2021.104148
15. YouTube: Figure status update - openai speech-to-speech reasoning (4/19/2024). https://www.youtube.com/watch?v=Sq1QZB5baNw

Generalized Authoring Tool for Computer Vision Machine Learning Application Deployments

Peter Stein[1,2(✉)], Jibinraj Antony[2], Simon Bergweiler[2], and Christian Schorr[1,3]

[1] University of Applied Sciences Kaiserslautern, Amerikastraße 1,
66482 Zweibrücken, Germany
[2] German Research Center for Artificial Intelligence (DFKI),
Trippstadter Strasse 122, 67663 Kaiserslautern, Germany
`peter.stein@dfki.de`
[3] German Research Center for Artificial Intelligence (DFKI),
Campus D 3.2, 66123 Saarbrücken, Germany

Abstract. Automated authoring enables simplified deployment of applications and services for complex use cases, especially in the field of machine learning. This paper presents the development and implementation of a specialized authoring tool that can be used for computer vision applications, enabling automated creation of machine learning services. The proposed authoring tool realizes a microservices architecture to facilitate the conversion and deployment of machine learning inference services, especially in image classification and object detection use cases. The authoring process addresses the interoperability issues commonly faced in machine learning frameworks, leveraging the Open Neural Network Exchange (ONNX) for model conversion into a standardized format. By encapsulating machine learning tools in containerized applications, this authoring tool offers a modular solution that can be easily adapted to various industrial applications. The developed authoring tool integrates the common machine learning frameworks PyTorch and TensorFlow, coupling DevOps methodologies such as CI/CD, ensuring a robust, maintainable, and user-friendly system that meets the growing needs of machine learning use cases in manufacturing.

Keywords: Machine Learning · Computer Vision · Containerization · Microservices · ONNX · PyTorch · TensorFlow · ML Authoring · ML Life cycle

1 Introduction

Machine Learning (ML) has revolutionized the manufacturing industry by enabling manufacturers to increase efficiency [12], reduce costs [13] and improve quality [8]. It is evident that effort-intensive and time-consuming manual tasks are being automated or facilitated by tools and services involving automation

K. Alexopoulos et al. (Eds.): ESAIM 2024, LNME, pp. 89–97, 2025.
https://doi.org/10.1007/978-3-031-86489-6_10

and ML. The sheer amount of data encountered every day can be trained and applied to a specific use case using diverse ML models.

For the training of the data, numerous frameworks in different programming languages are available today, such as PyTorch [15] or TensorFlow [19]. Each of these frameworks has its own format with its own characteristics and are generally not compatible with other frameworks. This causes problems, among other difficulties, in the area of sustainability and re-usability [6], as a new model has to be trained for each framework, making it difficult for researchers from different technical backgrounds to collaborate. Open Neural Network Exchange (ONNX)[5] offers a generalized platform enabling the interoperability for these ML models [7], but there lacks an automated solution for the deployment of the models to production environment. As the application of ML use cases increases, it is necessary to automate the application generation process, enabling efficient and error-free tools and services. This need gives opportunities for the development of automated authoring tools, capable of generating ML applications efficiently. Since no single framework is suitable for every problem in industrial applications, a combination of diverse tools and algorithms are often required to address different use cases.

The main contributions of this paper are the development of a generalized authoring tool, that can be utilized for ML deployments for computer vision-based use cases, and the evaluation of the authoring tool. This tool leverages a microservice architecture to facilitate the conversion of models and deployment of ML applications. It addresses the interoperability challenges commonly faced in ML by utilizing the ONNX framework for model conversion, into a unified format. By encapsulating ML applications in standalone lightweight executable containers, the tool offers a modular solution adaptable to various manufacturing scenarios in industrial applications. Additionally, the paper discusses the integration of PyTorch and TensorFlow frameworks to ensure a robust, maintainable, and user-friendly system that meets the growing needs of manufacturing. This paper is structured to provide a background and motivation for our work in Sect. 2, followed by a detailed description of the project's system architecture and methodology in Sect. 3, including insights into the integration with existing technologies like PyTorch, ONNX. Subsequent sections discuss the design and implementation of the tool, highlighting its capabilities in facilitating model conversion and deployment. We conclude the paper with a summary of our findings and potential future developments in Sect. 4.

2   Background

ML applications have significantly advanced industrial practices, especially in image classification and object detection for quality control and defect detection. Convolutional Neural Networks (CNNs) like MobileNet [9] or EfficientNet [18], as well as object detection models such as SSD [11] or Faster R-CNN [16], are widely used for their effectiveness in identifying patterns and anomalies [20].

Container-based solutions offer consistency and ease of scaling for ML models. TensorFlow Serving [2] and TorchServe [4] facilitate model deployment but can be resource-intensive and challenging for low-cost edge devices [10], due to their requirements for high-performance hardware to efficiently handle model serving. ONNX provides a more lightweight and versatile solution that facilitates model interoperability. However, the conversion process is frequently manual and complex [7].

Automated deployment tools are often used to prepare, configure and deploy ML applications to production environments. These tools often integrate continuous integration/continuous deployment (CI/CD) methods, which emphasizes the importance of automated testing and efficient deployment pipelines in the life cycle of ML applications.

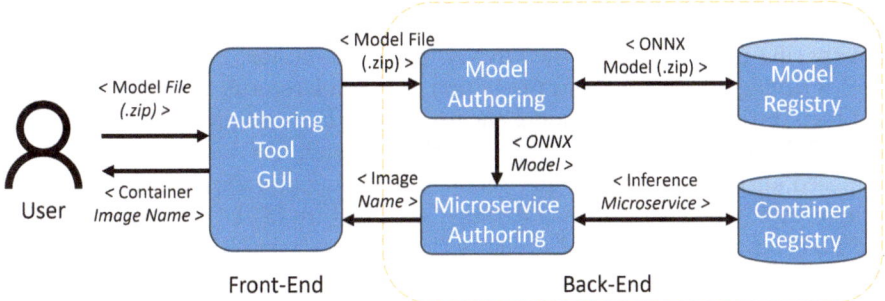

Fig. 1. Overview of the Automated ML Authoring Tool, illustrating the components involved in model and microservice authoring.

The concept of generalized automated deployment tools for ML models, particularly those involving deep learning models for image classification and object detection, is an emerging area in the field of artificial intelligence. While this idea is gaining traction, it remains relatively unexplored in the current academic literature. The development and implementation of such tools represents a significant step toward simplifying and rationalizing the deployment of models, thereby increasing their accessibility and usability in various application domains.

3 Automated Authoring Tool

In contrast to existing deployment frameworks like TensorFlow Serving and TorchServe, this paper presents a unique, generalized tool for preparation, configuration, and deployment of ML model. It integrates automated pre-processing, post-processing, model conversion, and encapsulation of a REST API within a standalone container. This tool utilizes a microservice-based orchestration architecture, containerization, and ONNX standardization to address interoperability, sustainability, and usability challenges when deploying computer vision-based ML applications. Specifically, this work automates the entire process from model input to containerized deployment, making integration and use in production environments easier. The tool's innovation lies in its holistic approach, which combines multiple automated steps into a single, seamless workflow. This provides a modular and adaptable solution suitable for a wide range of industrial applications.

Authoring tools are software components that prepare Assets for integration into the target Platform. These Assets can be data, models, algorithms or microservices. Traditionally, the preparation of microservices for integrating ML methods has been a manual task. In the context of this project, the visionary goal is to develop an automated authoring tool specifically for ML. This tool will support developers throughout the process of preparing and configuring execution environments for ML models and microservices. The automated authoring tool adopts a microservices architecture, with specialized components such as the user interface, conversion microservices each dedicated to converting models of a specific framework into the ONNX format, a database, inference microservices for image processing and prediction, and a microservice container creation and registration service for model encapsulation. Figure 1 provides the overview of the ML authoring tool developed in this research. As seen in the Fig. 1, the proposed authoring tool contains a model authoring and microservice authoring components. The Fig. 2 gives a detailed architecture of the ML authoring tool and its components. The tool's microservices interact via REST APIs.

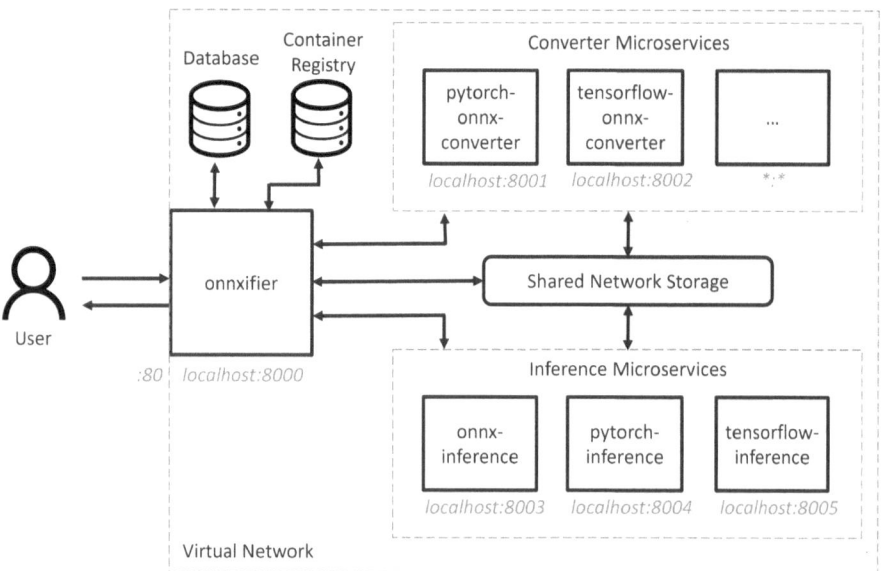

Fig. 2. Overview of the microservice architecture of the automated authoring tool, consisting of user interface, database, container registry, converter microservices, inferences microservices and a shared network storage.

3.1   Model Authoring

In the initial phase of the automated model authoring process, a user contributes a pre-trained ML model through the front-end component in a format depending on the framework the model has been trained. For instance, models trained in PyTorch are exported as entire models in .pth utilizing python's pickle module or the common *.pt* format. Similarly, TensorFlow models typically adopt the *SavedModel* format, which consists of a directory containing multiple files representing the neural network and its attribute values. To ensure uniformity, the user submits the model encapsulated in a .zip archive. The tool currently supports pre-trained ML models primarily focused on image classification and object detection tasks, with the possibility to incorporate additional model types in the future.

The archived model, in addition to the user-provided metadata, such as specific pre-processing parameters or the dataset the model was trained on, is then passed to a converter microservice. Given the diversity of formats prevalent in ML, multiple libraries come into play during this conversion process, as there is no singular library capable of handling all possible formats. PyTorch, for instance, has built-in capabilities for ONNX conversion, which can be easily integrated into the process. In the case of TensorFlow, model conversion is performed by the tf2onnx [3] library. In addition, there are general purpose conversion libraries such as ONNXMLTools that provide support for a wide range of ML frameworks, including TensorFlow, scikit-learn and Apple Core ML [1,14].

The result of this conversion is the authored ONNX model, which is stored in a configured network storage, as illustrated in Fig. 2. The storage and retrieval of these models are managed through an internal SQL Database. This converted model is then passed to the microservice authoring component for further processing.

3.2   Microservice Authoring

The microservice authoring module encapsulates an ML model in ONNX format, that has been generated in the model authoring stage, into a container image. This has REST API endpoints to perform the inference and thus the prediction. Within this authoring process, specific pre- and post-processing steps are integrated based on the framework and the use case.

In the pre-processing phase, an image is converted using the metadata specified by the user or required by the original framework, as shown in the Fig. 3. The image is first converted to a consistent format through operations such as resizing, cropping, and then normalized based on the mean and standard deviation of the training dataset. After this transformation process, the image is turned into a NumPy-array-based Tensor. Since most ML models are trained with open source datasets such as ImageNet, this step can be predefined and even preselected by the user.

The inference step within the microservice authoring process relies on the pre-processed Tensor from the previous step. For this, the ONNX Runtime module is used to perform the inference by generating an ONNX Inference Session.

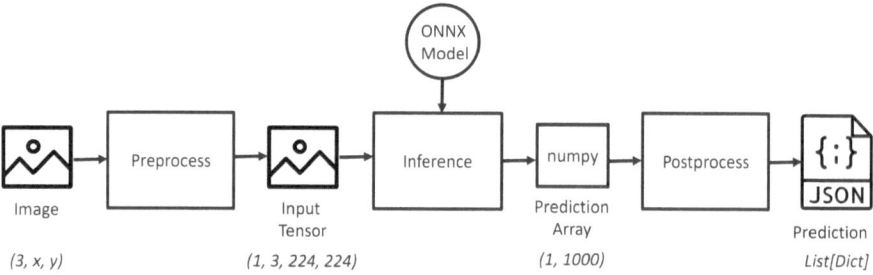

Fig. 3. The image pre-processing and post-processing pipeline. This illustrates an example process by which the original image is transformed to produce processed tensor data with the required dimensions. Inference is then performed using the authored ONNX model, and the resulting prediction tensor is transformed to a human-readable JSON format.

This session receives the processed Tensor and performs the prediction. Following the ONNX Inference session, the output is subjected to a further transformation in the post-processing step. Here, the raw results of the inference are transformed to a human-understandable JSON format. This JSON format also contains additional information, including the image class and the confidence probability. The generated microservices are pushed to a container registry with unique tag names. This final container image name, including the container registry URL and tags, is provided to the user after the authoring process. The user can use this information to configure the deployment on the target platform or their local environment.

The generated inference microservice follows a workflow as given in Fig. 4, which has been illustrated with an example use case of an image processing ML application packaged as Docker container.

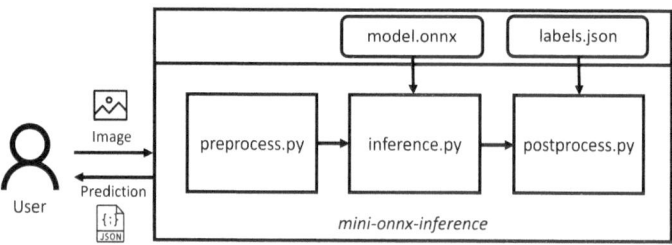

Fig. 4. Workflow of inference microservice packaged in Docker, generated using the ML authoring tool, illustrated with an image classification example, where a client sends an input image and receives a response in JSON format.

3.3   Container Image Registry

After successfully encapsulating ONNX models into ML applications as container images, these containers are pushed into a container image registry, such as the Docker Registry [17]. The container registry serves as a central repository for storing and managing container images, facilitating version control and ensuring accessibility. Users can push their containerized applications to the registry, making them readily available for deployment.

The container registry also aims to provide a secure and scalable storage solution that allows users to efficiently manage and share their container images. To deploy the encapsulated models, users can pull the images from the registry into their desired environments.

4   Conclusion and Outlook

The implementation of a generalized authoring tool for computer vision ML applications provides a comprehensive solution to the challenges of model conversion and deployment through automation. By leveraging a microservice architecture, the tool streamlines the process from user interaction through model conversion to containerized deployment, ensuring isolated and standardized execution across different frameworks. This modularity increases the tool's versatility, enabling the deployment of diverse ML applications and use cases.

While the tool currently supports image-based machine learning use cases, such as image classification and object detection, future extensions could broaden its applicability. Expanding support to include additional types of ML models, such as those used in natural language processing or audio analysis, would require ensuring that the REST API remains consistent and may involve introducing standardized interfaces to handle various data formats and model outputs.

Future work could explore the integration of additional frameworks in other programming languages, broadening the tool's applicability and facilitating its use across a wider range of machine learning tasks.

References

1. ONNXMLTools: A set of tools for working with onnx models. https://github.com/onnx/onnxmltools. Accessed 15 Dec 2023
2. TensorFlow Serving: A flexible, high-performance serving system for machine learning models. https://www.tensorflow.org/serving. Accessed 15 Dec 2023
3. tf2onnx: Convert tensorflow, keras, tensorflow.js and tflite models to onnx. https://github.com/onnx/tensorflow-onnx/. Accessed 10 Jan 2024
4. TorchServe: An open-source model serving library for pytorch. https://github.com/pytorch/serve. Accessed 15 Dec 2023
5. Onnx: Open neural network exchange (2019). https://onnx.ai. Accessed 15 Dec 2023
6. Araz, J.Y., et al.: Les houches guide to reusable ml models in lhc analyses (2024)

7. Bai, J., et al.: Onnx: Open neural network exchange format. arXiv preprint arXiv:1901.00069 (2019)
8. Chatterjee, S., Misbahuddin, M., Vamsi, P., Ahmed, M.H.: Power quality improvement and fault diagnosis of pv system by machine learning techniques. In: 2023 International Conference on Signal Processing, Computation, Electronics, Power and Telecommunication (IConSCEPT), pp. 1–6. IEEE (2023)
9. Howard, A.G., et al.: Mobilenets: Efficient convolutional neural networks for mobile vision applications. ArXiv **abs/1704.04861** (2017). https://api.semanticscholar.org/CorpusID:12670695
10. Lin, Q.: Smartlite: A dbms-based serving system for dnn inference in resource-constrained environments. Proc. VLDB Endow. **17**(3), 278–291 (2023)
11. Liu, W., et al.: Ssd: Single shot multibox detector. In: European Conference on Computer Vision (2015). https://api.semanticscholar.org/CorpusID:2141740
12. Matrenin, P., Antonenkov, D., Arestova, A.: Energy efficiency improvement of industrial enterprise based on machine learning electricity tariff forecasting. In: 2021 XV International Scientific-Technical Conference on Actual Problems Of Electronic Instrument Engineering (APEIE), pp. 185–189. IEEE (2021)
13. Osypanka, P., Nawrocki, P.: Resource usage cost optimization in cloud computing using machine learning. IEEE Trans. Cloud Comput. **10**(3), 2079–2089 (2020)
14. Pedregosa, F., et al.: Scikit-learn: machine learning in Python. J. Mach. Learn. Res. **12**, 2825–2830 (2011)
15. PyTorch Authors: PyTorch: An open source machine learning framework that accelerates the path from research prototyping to production deployment. PyTorch (2022). https://pytorch.org. Accessed 15 Dec 2023
16. Ren, S., He, K., Girshick, R.B., Sun, J.: Faster r-cnn: Towards real-time object detection with region proposal networks. IEEE Trans. Pattern Anal. Mach. Intell. **39**, 1137–1149 (2015). https://api.semanticscholar.org/CorpusID:10328909
17. Solomon Hykes: Docker registry: Distribution implementation for storing and distributing of container images and artifacts. https://hub.docker.com/_/registry. Accessed 15 Dec 2023
18. Tan, M., Le, Q.V.: Efficientnet: Rethinking model scaling for convolutional neural networks. ArXiv **abs/1905.11946** (2019). https://api.semanticscholar.org/CorpusID:167217261
19. TensorFlow Authors: TensorFlow: An open-source machine learning framework. TensorFlow (2022). https://www.tensorflow.org. Accessed 15 Dec 2023
20. Xia, B., Cao, J., Wang, C.: Ssim-net: Real-time pcb defect detection based on ssim and mobilenet-v3. In: 2019 2nd World conference on mechanical engineering and intelligent manufacturing (WCMEIM), pp. 756–759. IEEE (2019)

Collaborative Learning in Shared Production Environment Using Federated Image Classification

Vinit Hegiste[1(✉)], Tatjana Legler[1,2], and Martin Ruskowski[1,2]

[1] Chair of Machine Tools and Control Systems, RPTU Kaiserslautern-Landau,
Kaiserslautern, Germany
vinit.hegiste@rptu.de
[2] Innovative Factory Systems (IFS), German Research Center for Artificial
Intelligence (DFKI), Kaiserslautern, Germany

Abstract. The application of federated learning (FL) in industrial settings offers promising advancements in maintaining data privacy while collaboratively training machine learning models. This study focuses on the comparative analysis of federated image classification versus locally trained models within a shared production environment. Specifically, we explore the classification of windshields in truck cabins, which is a crucial task for quality inspection in manufacturing of trucks. Our research involves four clients, each producing different types of truck cabins and research based on FL process between them. Various deep learning architectures, including VGG19, ResNet50, InceptionNetv3, DenseNet-121, and EfficientNetv2-s, were evaluated under a FL framework implemented using the FLOWER framework. A custom plain averaging strategy was used for weight aggregation. The global model's performance was assessed using a combined test set from all clients and compared against models trained locally by individual clients. The results highlight the effectiveness of FL in enhancing model generalization and adaptability to new product variations in industrial applications, promoting its adoption for collaborative quality inspection tasks.

Keywords: Federated Learning · Image Classification · Quality Inspection · Deep Learning · Industrial Applications

1 Introduction

The quality of the dataset is pivotal in training machine learning models. High-quality datasets lead to the development of robust models that perform effectively across a range of applications [4]. Federated learning (FL), a distinct paradigm of machine learning, facilitates the training of a cohesive model through the collaborative efforts of multiple clients. This approach involves the aggregation of model weights from each participant, ensuring that the training data remains on the local servers, and enhancing the model's ability to perform

© The Author(s) 2025
K. Alexopoulos et al. (Eds.): ESAIM 2024, LNME, pp. 98–106, 2025.
https://doi.org/10.1007/978-3-031-86489-6_11

in unfamiliar testing environments [16]. Recent years have witnessed a surge in FL applications, driven by the growing demand for data privacy and the need for collaborative solutions across industries [21,22]. Despite its increasing popularity, the application of FL in visual tasks within the manufacturing sector for custom datasets does not have much research compared to the FL algorithms and architectures tested on IID (independent and identically distributed) datasets such as CIFAR-10 and MNIST [9,21]. Furthermore, there is a notable scarcity of studies comparing models trained via FL with those trained using traditional, local datasets. Such comparisons are crucial in the commercial sector, as they highlight the differences in performance on test datasets between locally trained models and those trained through a federated approach [3]. This contrast not only showcases the effectiveness of FL in enhancing model generalization across diverse datasets but also encourages more companies to engage in FL initiatives [15].

This research focuses on the comparative analysis of federated image classification within a shared production environment. Our study involves four clients, each producing different types of truck cabins, with or without windshields. We examine the efficacy of various deep learning architectures in a FL setting and evaluate two distinct strategies for weight selection in federated models. This investigation aims to shed light on the optimal configurations and strategies that enhance performance in FL applications, particularly in industrial settings.

2 Related Work

The adoption of FL in industrial applications, particularly for quality inspection and predictive maintenance, has garnered significant interest. This interest is driven by FL's ability to train models collaboratively without compromising data privacy. [5] explored failure prediction using FL on production lines, illustrating the efficacy of FL in real-world scenarios. [9] furthered this research by developing federated object detection algorithms for quality inspection tasks in manufacturing environments.

Introduced by McMahan et al. [16], Federated Averaging (FedAvg) has become a foundational algorithm in FL, enabling multiple devices to collaboratively train a model while maintaining data localization and privacy. However, there has been limited exploration into comparing different deep learning architectures within FL frameworks like FedAvg to assess their impact on model performance [8,13]. Evaluating these architectures in an FL context is particularly important, as highlighted by [1], given their widespread use in image classification and the varying complexity they offer. Studies such as those by [8,9], and [10] have begun addressing this gap by analyzing the performance differences between federated and centralized models in industrial settings. Additionally, [14] explored strategies for integrating new clients into FL networks, enhancing performance in dynamic environments. Further work by [7] demonstrated FL's ensemble capabilities, showing improved object detection in previously unseen scenarios, further supporting the case for FL's application in complex industrial contexts.

3   Implementation

This section outlines the FL architecture and framework utilized, the distribution and characteristics of the dataset for experimentation, and the external test dataset employed for evaluating the globally generated models against the locally trained client models.

3.1   Federated Learning Framework

Several FL frameworks facilitate research by simplifying the integration and testing processes. Notable frameworks include TensorFlow Federated [12], PySyft [20], and FLOWER [2]. Among these, FLOWER is chosen for its ease of integration and effectiveness in research-oriented applications. FLOWER is a flexible and user-friendly framework that supports various experimental setups.

In our FL setup, all clients shared the same deep learning architecture and hyperparameters to ensure consistency across the federated learning process. The architectures varied, but the hyperparameters were kept constant: batch size=16, optimizer=SGD with momentum=0.8, learning rate=0.001, loss function=cross-entropy, and image size=300. Deep learning architectures-EfficientNetv2 (small) [19], VGG19 [17], ResNet50 [6], DenseNet-121 [11], and InceptionNetv3 [18] were selected for their proven performance in image classification tasks and their varying complexities in terms of trainable parameters. This selection allows for a comprehensive analysis of how different architectures impact the effectiveness of federated learning in handling diverse and complex data scenarios, particularly in an industrial setting. The FL strategy employed was plain averaging of model weights for the global federated model, customized within the FLOWER framework for each architecture as mentioned in Table 1.

3.2   Dataset

The primary scenario for this research involves detecting the presence of a windshield in truck cabins as a quality inspection application. The dataset comprises four clients, each identified by the color of their cabins: Blue, Green, Orange, and Red. Each client's dataset includes two labels: 'No_windshield' and 'With_windshield', as illustrated in Fig. 1. The total data distribution can be referred to in Fig. 2. An external test dataset was also developed to challenge the robustness of local models under FL paradigms. This dataset includes cabins of different colors (gray and purple) and features a novel type of windshield, depicted on the right side of Fig. 1.

3.3   Experimental Procedure

We began with an FL architecture where all clients shared the same deep learning architecture and hyperparameters. The architectures were varied, but the hyperparameters were kept constant: batch size=16, optimizer=SGD with momentum=0.8, learning rate=0.001, loss function=cross-entropy, and image size=300,

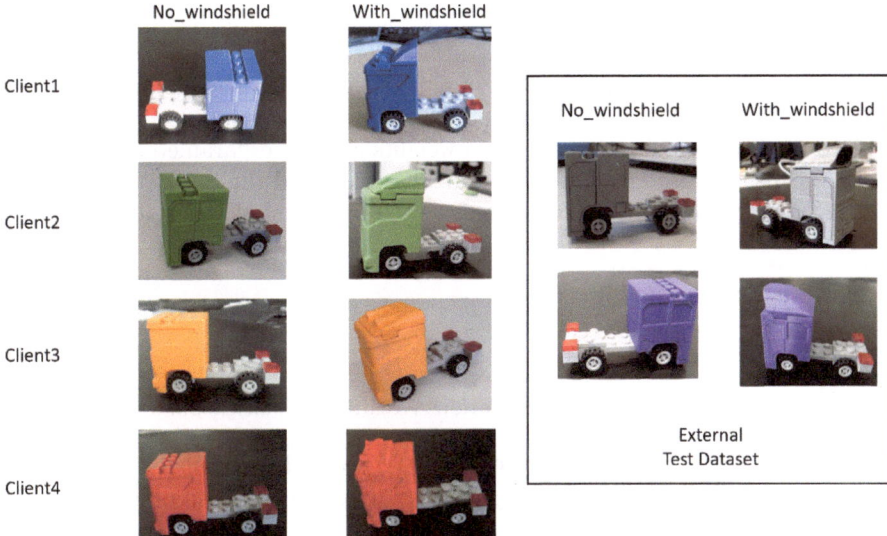

Fig. 1. Local Client dataset for each client in FL (left) and external test dataset with 2 Colored Cabins along with a novel Cabin type which none of the clients have ever seen before (right)

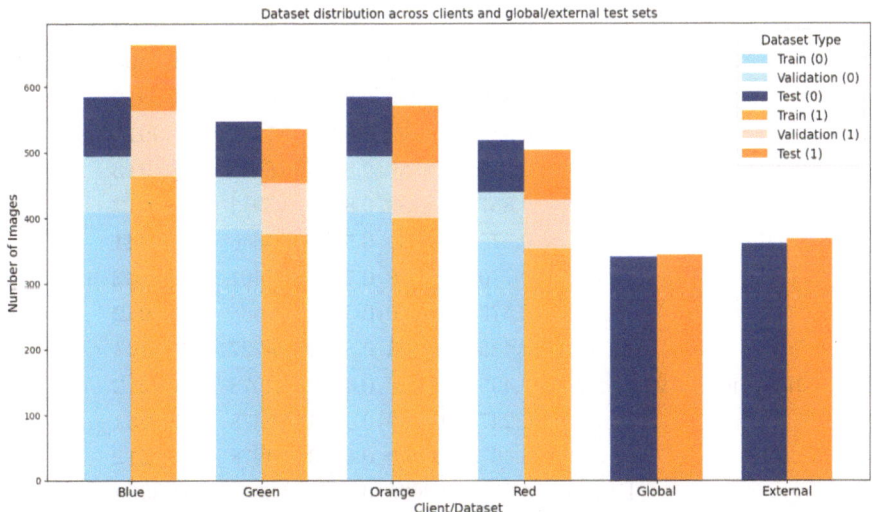

Fig. 2. Distribution across clients and global/external testsets. 'No_windshield' is represented as 0 and 'With_windshield' is represented as 1. The Global dataset is a combination of the test set of all 4 clients.

using the final weights from each local epoch. The FL strategy employed was custom plain averaging of model weights for the global federated model, customized within the FLOWER framework for different deep learning architectures as mentioned in Table 1. To expedite the testing process, a global test set was created by amalgamating the test sets of all clients. The global model was evaluated against this global test set following each communication round to achieve the best combination of global weights and hyperparameters. This approach was used to evaluate the final global models and the best locally trained models, demonstrating the effectiveness of the federated global model.

4 Results and Discussion

Table 1. Performance Metrics for Various Architectures on Global Test Dataset

Architecture	Metric	Client1	Client2	Client3	Client4	Global Model
DenseNet-121	Accuracy	0.7438	0.8287	0.7818	0.6823	**0.9971**
	Precision	0.8207	0.8495	0.8210	0.6837	**0.9971**
	Recall	0.7438	0.8287	0.7818	0.6823	**0.9971**
	F1 score	0.7278	0.8262	0.7752	0.6818	**0.9971**
EfficientNetv2	Accuracy	0.6633	0.7277	0.6281	0.6310	**0.9898**
	Precision	0.7694	0.7695	0.7580	0.6320	**0.9899**
	Recall	0.6633	0.7277	0.6281	0.6310	**0.9898**
	F1 score	0.6272	0.7170	0.5756	0.6302	**0.9898**
VGG19	Accuracy	0.8594	0.8389	0.8873	0.6428	**0.9912**
	Precision	0.8827	0.8399	0.8923	0.7920	**0.9913**
	Recall	0.8594	0.8389	0.8873	0.6428	**0.9913**
	F1 score	0.8574	0.8389	0.8869	0.5913	**0.9912**
ResNet50	Accuracy	0.5754	0.7130	0.5007	0.5666	**0.9941**
	Precision	0.7709	0.8148	0.7507	0.7401	**0.9942**
	Recall	0.5754	0.7130	0.5007	0.5666	**0.9942**
	F1 score	0.4834	0.6883	0.3374	0.4727	**0.9941**
InceptionNetv3	Accuracy	0.4978	0.4890	0.4890	0.4978	**0.5212**
	Precision	0.2478	0.4569	0.4704	0.2478	**0.6231**
	Recall	0.4978	0.4890	0.4890	0.4978	**0.5212**
	F1 score	0.3309	0.3636	0.3829	0.3309	**0.3998**

After extensive experimentation with different communication rounds (CRs) and epochs, the optimal global federated model was achieved using 5 local epochs and 15 CRs. This section presents the performance metrics of various deep learning architectures, comparing both individual client models and the federated global model on the global test dataset. Table 1 outlines the Accuracy, Precision, Recall, and F1 scores for each model. The federated global model consistently outperformed individual client models across all architectures. For instance, DenseNet-121 achieved a global model F1 score of 0.9971, significantly higher than the individual client F1 scores, which ranged from 0.6818 to 0.8262. Similar trends were observed with other architectures, where the federated model demonstrated superior performance, underlining the effectiveness of federated learning (FL) in improving model generalization.

Table 2. F1 Scores on External Test Dataset (Gray and Purple Cabins) for Centralized vs. Federated Training

Training Type	DenseNet	EfficientNet	VGG	ResNet	InceptionNet
Centralized	0.8569	0.6102	0.9100	0.7654	0.4967
Federated	**0.9821**	**0.9876**	**0.9586**	**0.9917**	0.4394

To further evaluate the robustness of these models, we tested both centralized and federated models on an external test dataset consisting of gray and purple cabins with an unseen windshield type. This scenario simulates a real-world use case where a company integrates a new windshield type into its manufacturing process, and the goal is to assess how well pretrained models can handle such unseen data. Table 2 presents the F1 scores for both centralized and federated models on this external test dataset. The results indicate that federated models generally outperform their centralized counterparts, particularly with DenseNet, EfficientNet, VGG, and ResNet architectures, achieving F1 scores of 0.9821, 0.9876, 0.9586, and 0.9917, respectively. This demonstrates the superior robustness and generalization capability of federated models when exposed to unseen data, highlighting their potential for real-world industrial applications where new components or product variations are frequently introduced.

In summary, the results highlight that federated learning not only enhances model performance on combined datasets but also significantly improves the model's ability to generalize to new, unseen scenarios. The VGG19 model has approximately 143.67 million parameters, ResNet50 has around 25.56 million parameters, DenseNet-121 has about 7.98 million parameters, EfficientNetv2 has approximately 21.55 million parameters, and InceptionNetv3 has about 23.85 million parameters. Despite having a relatively lower number of trainable parameters, DenseNet's federated global model achieved near-perfect performance metrics, demonstrating its efficiency and suitability for resource-constrained environments. These findings emphasize the efficacy of federated learning in industrial settings, where data privacy and the ability to adapt to new conditions are paramount.

5 Conclusion and Outlook

This study explored the application of Federated Learning (FL) for image classification within a shared production environment, focusing on classifying windshields in truck cabins. We evaluated the performance of several deep learning architectures, comparing models trained locally by individual clients with a global model obtained through FL using a custom plain averaging strategy. The experimental results indicate that FL significantly enhances model performance across all tested architectures, consistently achieving higher accuracy, precision, recall, and F1 scores compared to individual client models. This portrays FL's potential to create robust and generalized models by aggregating knowledge from multiple sources while preserving data privacy. Furthermore, testing on an external dataset with unseen windshield types demonstrated the adaptability and robustness of federated models in handling unforeseen data. Considering both performance metrics and the total number of trainable parameters, DenseNet-121 emerged as the most suitable architecture, offering near-perfect performance with fewer trainable parameters, making it both efficient and resource-friendly. In contrast, InceptionNetv3 consistently underperformed, indicating its unsuitability for this task.

The results also highlight the limitations of local models trained on isolated datasets, which perform poorly in comparison. FL addresses this by enabling a superior global model without data sharing, critical in industrial applications where data privacy is paramount. Future work will extend this approach to other quality inspection tasks and explore advanced FL strategies, such as differential privacy and secure multi-party computation, to enhance data security. Additionally, integrating FL with real-time industrial systems for continuous learning and adaptation to new production scenarios will be investigated.

Acknowledgment. This work was funded by the Carl Zeiss Stiftung, Germany under the Sustainable Embedded AI project (P2021-02-009).

References

1. Alzubaidi, L., Zhang, J., Humaidi, A., et al.: Review of deep learning: concepts, CNN architectures, challenges, applications, future directions. J. Big Data **8**(53), 1–74 (2021)
2. Beutel, D.J., Topal, T., Mathur, A., Qiu, X., Parcollet, T., Lane, N.D.: Flower: A friendly federated learning research framework (2020). https://flower.dev/
3. Bonawitz, K., et al.: Practical secure aggregation for privacy-preserving machine learning. In: Proceedings of the 2017 ACM SIGSAC Conference on Computer and Communications Security, pp. 1175–1191. ACM (2017)
4. Chen, H., Chen, J., Ding, J.: Data evaluation and enhancement for quality improvement of machine learning. IEEE Trans. Reliab. **70**, 831–847 (2021)
5. Ge, N., Li, G., Zhang, L., Liu, Y.: Failure prediction in production line based on federated learning: an empirical study. J. Intell. Manuf. **33**, 2277–2294 (2021)
6. He, K., Zhang, X., Ren, S., Sun, J.: Deep residual learning for image recognition. In: CVPR, pp. 770–778 (2016)
7. Hegiste, V., Legler, T., Ruskowski, M.: Federated ensemble yolov5: enhancing object detection via federated learning. In: Proceedings of the International Conference on Machine Learning and Privacy, pp. 45–60 (2021)

8. Hegiste, V., Legler, T., Ruskowski, M.: Application of federated machine learning in manufacturing. In: 2022 International Conference on Industry 4.0 Technology (I4Tech), pp. 1–8 (2022)

9. Hegiste, V., Legler, T., Ruskowski, M.: Federated object detection for quality inspection in shared production. In: 2023 Eighth International Conference on Fog and Mobile Edge Computing (FMEC), pp. 151–158 (2023)

10. Hegiste, V., Walunj, S., Antony, J., Legler, T., Ruskowski, M.: Enhancing object detection with hybrid dataset in manufacturing environments: Comparing federated learning to conventional techniques (May 2024), presented at the International Conference on Innovative Engineering Sciences and Technological Research (ICIESTR-2024)

11. Huang, G., Liu, Z., Maaten, L.V.D., Weinberger, K.Q.: Densely connected convolutional networks. In: CVPR pp. 4700–4708 (2017)

12. Inc., G.: Tensorflow federated: Machine learning on decentralized data. Software available from tensorflow.org (2019). https://www.tensorflow.org/federated

13. Islam, F., Raihan, A.S., Ahmed, I.: Applications of federated learning in manufacturing: Identifying the challenges and exploring the future directions with industry 4.0 and 5.0 visions. arXiv preprint arXiv:2302.13514 (2023)

14. Legler, T., Hegiste, V., Ruskowski, M.: Mapping of newcomer clients in federated learning based on activation strength. J. Manufact. Federated Learn. **1**(1), 100–110 (2021)

15. Li, T., Sahu, A.K., Talwalkar, A., Smith, V.: Federated learning: challenges, methods, and future directions. IEEE Signal Process. Mag. **37**(3), 50–60 (2020)

16. McMahan, H.B., Moore, E., Ramage, D., Hampson, S., y Arcas, B.A.: Communication-efficient learning of deep networks from decentralized data. In: AISTATS, pp. 1273–1282 (2017)

17. Simonyan, K., Zisserman, A.: Very deep convolutional networks for large-scale image recognition. In: ICLR (2015)

18. Szegedy, C., Vanhoucke, V., Ioffe, S., Shlens, J., Wojna, Z.: Rethinking the inception architecture for computer vision. In: CVPR, pp. 2818–2826 (2016)

19. Tan, M., Le, Q.V.: Efficientnet: Rethinking model scaling for convolutional neural networks. In: ICML, pp. 6105–6114 (2019)

20. Trask, A., Mancuso, J., Ryffel, T., Rueckert, D.: Pysyft: A library for encrypted, privacy preserving machine learning. In: 31st Conference on Neural Information Processing Systems (NIPS 2018) (2018)

21. Wen, J., Zhang, Z., Lan, Y., Jin, W.: A survey on federated learning: challenges and applications. Int. J. Mach. Learn. Cybern. **14**, 513–535 (2023)

22. Yang, Q., Liu, Y., Chen, T., Tong, Y.: Federated machine learning: concept and applications. ACM Trans. Intell. Syst. Technol. (TIST) **10**(2), 12 (2019)

An OPIS-Based Knowledge Engineering Framework for Collaborative Robotics

Elisa Foderaro[1,2], Alessandro Umbrico[2], and Andrea Orlandini[2(✉)]

[1] University of Genoa, Genoa, Italy
[2] ISTC, National Research Council of Italy, Rome, Italy
andrea.orlandini@istc.cnr.it

Abstract. Artificial Intelligence can allow cobots to work autonomously, perceiving and understanding the environment, planning tasks, and properly acting to achieve production goals. However, effective deployment of AI technologies in real plants is not straightforward. This paper proposes the extension of SOHO (Sharework Ontology for Human-Robot Collaboration), an ontology for Human-Robot Collaboration, with OPIS concepts, a language for describing manufacturing processes. We also present a new extended implementation of TENANT, a Knowledge Engineering tool based on SOHO and OPIS. We show its suitability in a realistic collaborative scenario with a human and a robot operator cooperating to perform an assembly process.

Keywords: A.I. · Knowledge Engineering · Collaborative Robotics

1 Introduction

Achieving an effective deployment of collaborative robots in manufacturing is a complex task. According to [4], different latent dimensions can be considered in evaluating a human-robot collaboration (HRC) framework. The first is autonomy, i.e. the ability to sense, plan, and act according to the task. High level autonomy requires control systems able to efficiently implement production tasks while adapting to changes in the environment. Artificial Intelligence (AI) plays a key role in this context, enabling robots to operate autonomously, perceiving and understanding the working environment, planning their tasks, and properly acting to achieve specific goals. However, effective deployment of AI technologies in real industrial environments is not straightforward. In particular, building proper input specifications for an AI framework is usually complex. Indeed, there is a lack of a generally accepted modelling methodology and this potentially leads to large efforts to define suitable robot control specifications. Moreover, there is still a clear need for knowledge engineering tools to facilitate communication and interaction between AI and Robotics engineers as well as with domain experts. It is essential to develop tools that can help these experts and enable seamless collaboration. Recently, a software tool called TENANT [3] was proposed to assist production engineers in defining goals, tasks, and

© The Author(s) 2025
K. Alexopoulos et al. (Eds.): ESAIM 2024, LNME, pp. 107–116, 2025.
https://doi.org/10.1007/978-3-031-86489-6_12

operational constraints to provide the automatic generation of task planning specifications for robot control in collaborative scenarios. The representation framework behind TENANT relies on SOHO (Sharework Ontology for Human-Robot Collaboration) [15], a state-of-the-art domain ontology that provides a "standard" semantics to represent production-related knowledge and, therefore, interpret production engineers' input. It supports a contextual and hierarchical structuring of knowledge. This structure is encapsulated by TENANT to collect production-related information, store knowledge, and implement an automatic task planning model generation. However, it is clear the need to support also a complete (abstract) representation of specific production knowledge for a whole manufacturing shop-floor, and not only concerning a single-robot production process. In particular, a tool was needed to model also large-scale manufacturing systems, while maintaining the possibility to selectively and dynamically vary the level of detail at which the system can be observed. In this regard, OPIS [14] is a framework used in real manufacturing scenarios particularly relevant to our objective. It proposes a well-structured formalism describing production resources and operations carried out within a manufacturing system. Our goal is to propose a process-independent and complete knowledge engineering approach for the synthesis of AI planning models. To this aim, SOHO and OPIS are used as the knowledge structure in TENANT. Introducing OPIS concepts is also crucial to maintain consistency between SOHO and TENANT to tightly integrate a Knowledge Base (KB) built upon SOHO and to further facilitate task planning models definition. Linking an ontology can contribute in reducing the risk of user errors. Indeed, the ontology's reasoning mechanisms can be exploited to introduce semi-automatic steps in TENANT pipeline. The contributions of the paper are: definition of an extension of SOHO with OPIS concepts, presentation of a new TENANT implementation based on SOHO+OPIS, and showing its suitability in a realistic collaborative scenario. The paper is organized as follows: Sect. 2 provides an overview on Knowledge Engineering and OPIS; Sect. 3 discusses the use of ontology in manufacturing and the pursued approach; Sect. 4 describes the representation formalism and the modeling process; Sect. 5 shows the functioning of TENANT in a collaborative use case; Sect. 6 draws some conclusions.

2   Knowledge Engineering for Production Planning

Knowledge Engineering (KE) includes methods and tools aimed at improving the process of acquiring, using, and implementing engineering knowledge and automation. The definition, validation and reuse of knowledge within automation processes are central to engineering, contributing to reduction in costs and product development time [12]. The reuse of knowledge indeed decreases the engineering resources required relieving engineers from non-value-adding activities. Several works have investigated the design and use of KE tools to formally describe complex products and processes, and reduce the costs of automation processes [1,5,7,9]. These works address important challenges of KE but do not support production engineers to model production dynamics and to automatically synthesize planning models suitable to control autonomous robots at both the "production unit" level (e.g., within a collaborative cell [3]) and the shop-floor level. Other works addressed the automatic synthesizing of scheduling plans [11], or their manual creation, editing and management [8] through graphical user interfaces. However, these approaches do not primarily focus on the initial phase of modelling the environment and its dynamics, relying instead on the automatic retrieval of requirements or on the manual creation of plans. In [13], KE and Automated Planning are integrated to support engineering of human-robot interaction dynamics, which are then translated into a planning model for the actual coordination of robot parts. Although in a different domain, this work pursues a similar objective but we are interested in addressing the problem of modeling large-scale manufacturing systems, and OPIS is relevant to our discussion. OPIS [14] is a general framework for modeling manufacturing systems, with a focus on large-scale ones. It is particularly relevant as it provides an extensible set of modeling primitives that emphasize the development and use of hierarchical models of manufacturing processes and required resources.

Utilizing OPIS in describing manufacturing processes offers several advantages. First, it allows for the detailed specification of the myriad of constraints that govern manufacturing environments, such as resource limitations or production deadlines. In doing so, it ensures that models accurately reflect the realities of the manufacturing process. Secondly, OPIS supports the development and use of hierarchical models, providing a clear and organized representation of the manufacturing environment. This hierarchical structure is essential as manufacturing processes and resources are often structured hierarchically, with multiple levels of processes and sub-processes, each requiring specific resources. Additionally, OPIS provides an extensible set of modeling primitives and captures the dynamic nature of manufacturing environments, where processes and constraints can change over time. This extensibility and flexibility ensure long-term applicability and utility. These capabilities are crucial for effective production management and control decision-making, as they enable the identification and resolution of potential bottlenecks and inefficiencies as well as enabling more effective planning and scheduling. A model is specified in terms of five basic types of entities. *Resources* describe of the various physical/logical entities required to perform manufacturing activities. OPIS proposes a hierarchical description of resources, that can be organized into *Aggregated Resources*, which can be either aggregated entities composed of several (simpler) parts/objects, or more abstract sets of resources. *Operations* describe specific activities performed within the manufacturing system. Operations are hierarchically organized to create descriptions of manufacturing processes. OPIS distinguishes two forms of process abstractions: *conjunctive abstractions*: operations decomposed into a sequence of (sub-)operations at a lower level of the hierarchy, *disjunctive abstractions*: operations decomposed into a set of alternative (sub-)operations at a lower level in the hierarchy. In this sense, manufacturing processes are just sets of hierarchically organized operations. *Products* describe materials produced by manufacturing systems, either as final outputs or as input materials for more complex objects (e.g., semi-worked pieces to be used in production sub-processes). Product descriptions are organized hierarchically into *Product Families*, representing sets of products that share commonalities in their manufacturing processes. *Demands* describe obligations for product delivery and specify requests for quantities of specific products to be satisfied, with their production constraints, e.g. time constraints or client-dependent priorities. *Production units* represent collections of products manufactured together and created in response to product demands.

3   Semantic-Aided Definition of Production Knowledge

Describing procedures, capabilities of working entities, and their possible interactions to support (agile) production objectives is challenging. The description should take into account different perspectives (e.g., the single "local perspective" of each acting entity and the "global perspective" of the whole production system) and contextualize production dynamics (e.g., operative constraints, production goals) to the different features and skills of the involved agents (e.g., human workers, machines, autonomous robots). SOHO is structured in 3 main contexts, each considering a specific perspective. The *Environment Context* characterizes the physical objects composing the modeled production environment and the general properties that can be observed. The *Behavior Context* characterizes the behavior of the entities that can actively participate in the production processes, both in terms of low-level operations (and the capabilities required to perform them) and different levels of collaboration. The *Production Context* defines the production requirements and the tasks that should be performed to achieve desired production goals. The description of a production process follows a hierarchical, task-oriented approach. SOHO is a domain ontology specifically designed for characterizing production dynamics within collaborative cells. In contrast, OPIS is focused on modeling large-scale systems, e.g. an entire shop floor, but it does not explicitly support the requirements of HRC scenarios. Therefore, we propose the extension of SOHO including some key concepts from OPIS. We can then develop a more complete and versatile ontology to model the entire manufacturing process and support HRC scenarios as well. This can also support knowledge engineering tools like TENANT.

   First, we augment the knowledge definition of resources in SOHO by integrating `Aggregated Resources` from OPIS to encourage granularity of knowledge. Also, `Workpiece-type Resources`, present in SOHO, are replaced by OPIS `Products`, which basically represent the same concept. However, by introducing Products as a separate concept in SOHO, we can incorporate also the hierarchy based on product families from OPIS. This addition will be particularly useful in future versions of TENANT, as it will naturally facilitate the reusing of knowledge about manufacturing processes. Regarding the definition of `Production Processes`, SOHO provided a greater level of detail compared to OPIS, where all types of tasks are simply defined as *Operations*. In SOHO, the ontological concept `ProductionGoal` defines general goals to be achieved by the execution of the `ProductionMethod`. Each `ProductionMethod` is made of several hierarchically organized `ProductionTask` and is associated with several `ProductionNorm` constraining the task execution. SOHO distinguishes between `ComplexTask` and `SimpleTask`. `ComplexTask` describes an operation resulting from the composition of other tasks. This is used to characterize hierarchical decomposition of production tasks. A `ComplexTask` can be associated with either `SimpleTask` or other `ComplexTask`. Inspired by OPIS, we differentiate between two types of `ComplexTask`: `Conjunctive` or `Disjunctive`; a `SimpleTask` represents a leaf of the hierarchical structure and describes primitive production operations executed by performing a series of `Function`. Finally, according to [6], in SOHO the execution of a `HRCTask` can entail four different collaboration modalities between a human and a robot: (i) *Independent*, human and robot perform their tasks on different work-pieces without collaboration; (ii) *Simultaneous*, human and robot perform distinct tasks on the same work-piece at the same time, still without physical contact; (iii) *Supportive*, human and robot perform the same task on the same work-piece, working simultaneously and cooperatively on the same task. (iv) *Synchronous*, human and robot complete sequential tasks on the same work-piece, operating consecutively without any physical contact. The concept `HRCTask` is thus further specialized into four types of tasks. `IndependentTask` are implemented by a single `Function` that can be performed by a `HumanWorker` or by a `Cobot`. A human/robot performs the function independently from the other. `SimultaneousTask` are implemented by precisely two instances of `Function`, one function performed by a `Worker` (i.e., `HumanFunction`), another performed by a `Cobot` (i.e., `RobotFunction`). In this case, the human and the robot work on the same `WorkPiece` performing two different functions that can be carried out without any specific constraint. `SynchronousTask` are implemented by a `RobotFunction` and a `HumanFunction`. The pattern in this case forces the human and the robot to perform these two functions following a strict sequential order (this type of task is associated with the production norm `SequentialExecution`). `SupportiveTask` are implemented by a `RobotFunction` and a `HumanFunction`. It forces the human and the robot to perform two functions in parallel (this type of task is associated with the production norm `ParallelExecution`) achieving the highest level of collaboration. The basic structure of SOHO is then preserved with the only introduction of conjunctive and disjunctive abstractions in the definition of complex tasks. Finally, `Demands` are included in SOHO with the same specifications as in OPIS, while `Production Units` were not considered.

4   Knowledge Definition with TENANT

To be compliant with the knowledge defined in the SOHO extension, a new implementation of TENANT was developed. In this new version of TENANT, the knowledge base definition process consists of several modeling steps as shown in Fig. 1. The first step requires users to describe the physical configuration of a production environment and the objects that belong to it i.e., resources. Each resource is characterized by a name, a textual description, a capacity, and a type, which can be selected from some predefined types. In addition, a resource may belong to an aggregated resource. Each aggregated resource is characterized by a name and may belong to another aggregated resource, forming a hierarchy with multiple abstraction levels.

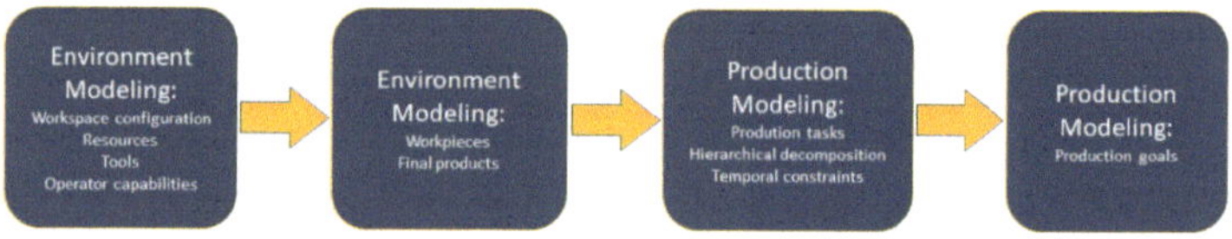

HRC acting entities, i.e., robots and human workers (agents), are resources that actively participate in the

Fig. 1. Production modeling pipeline of TENANT

collaborative production process. They are characterized, in addition to the properties of general resources, by a list of operations that they can perform according to their structure and skills, i.e. capabilities. At this stage, users must also specify these technical capabilities, each characterized by a name. The second step in the pipeline requires users to describe the products manufactured by the production system, both as final products and as intermediate ones. Products are characterized by a name and may be part of a product family. Unlike aggregated resources, the hierarchy of products is restricted to one level, i.e., a product can be part of a product family but each product family is characterized only by its name and cannot belong to another product family. However, this constraint can be easily removed if needed. Then, users must define processes in terms of tasks and operational requirements. The description of production processes follows a hierarchical, task-oriented approach. The goal of a manufacturing process is to produce a product by performing a series of operations (i.e., tasks). Thus, each product is associated with at least one process, while a process always refers to only one product. Moreover, each process is composed of several tasks and is associated with a set of relationships that impose "constraints" on the execution of such tasks. As a last step, users specify the production goals to be achieved, i.e. demands to state the requests for the shop floor and define the general production goals. The tool is implemented as a web-based application that provides the user with an interactive graphical environment for a step-by-step definition of the information characterizing the specific production environment. The tool is publicly available on GitHub[1].

5   TENANT at Work

An assembly scenario inspired by a real plant [2] and requiring a human and a collaborative robot to assemble a mosaic is considered here to assess TENANT and validate its functions.

---

[1] https://github.com/pstlab/TENANT_sa.git.

The shape of a "mosaic" consists of 5 rows and 10 columns to be filled with cubes of different colors. Each cell is identified by a letter for the column and a number for the row. In the end, the letters SW are formed, with the letter S made of orange cubes, the letter W made of white cubes, and the background made of blue cubes. The assembly process is also subject to some constraints: orange cubes can only be handled by the robot, white cubes can only be handled by the operator, and blue cubes can be managed by both. Additionally, there are some precedence constraints: Row 3 should start after the end of Row 1, Row 4 should start after the end of Row 1 and Row 2, Row 5 should start after the end of Row 1, Row 2 and Row 3.

The knowledge to represent this scenario was defined through TENANT and stored in a DB. The first step in the pipeline involves describing the physical configuration of the mosaic scenario, i.e. the objects needed in the assembly process. Figure 2 shows the resources page after adding all relevant resources. In this scenario, the "pick and place" action is the only relevant capability for the acting entities, i.e. a human worker and a cobot. As a consequence, it is the only capability added when defining the agents, as shown in Fig. 3(a), 3(b). Then, the products are defined by their names. Both the final product (the Mosaic) and the intermediate ones (the individual cells) were entered.

Fig. 2. TENANT interface.

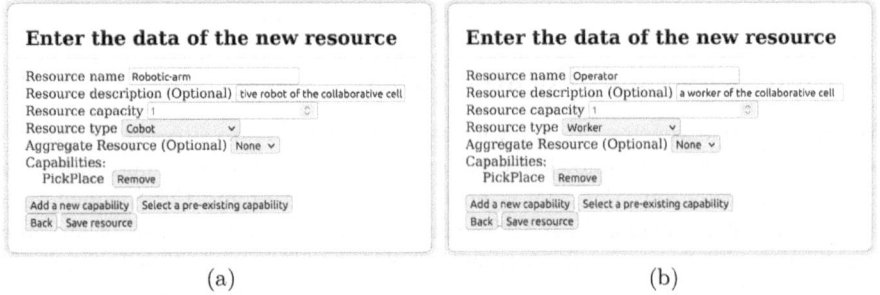

(a)   (b)

Fig. 3. TENANT user interface: information for (a) a cobot and (b) a human worker

The next step in the pipeline is the definition of the process in terms of the tasks to be achieved. This is the most time-consuming step and the one that would benefit the most from integrating TENANT with a KB.

Figure 4 shows an intermediate step in the definition of the process. The top-level task called *Root*, is a *Conjuctive Task* that is decomposed into five tasks, each defining the operations needed to complete a row in the mosaic. The *Root* task is important as it allows setting the precedence constraints of the require-ments. Figure 5 shows how they are added, at the end of the new process page.

Fig. 4. Intermediate Process definition

The first 3 cells of the mosaic, i.e. A1, B1, and C1, consist of orange cubes that can only be moved by the robot. For this reason, they are repre-sented as *Simple Task*s in the Independent modality and are associated with a single function of pick and place performed by the robot. The target products of these functions are the cells themselves, while the resources required are sim-ply the orange cubes. A similar description applies to the cells that need to be filled with white cubes, e.g. E1.

On the other hand, the cells that need to be filled with blue cubes, e.g. D1, are defined in a slightly more complex way. Since the blue cubes can be moved by both the human operator and the robot, the user must first add a disjunctive task that denotes the operation of assem-

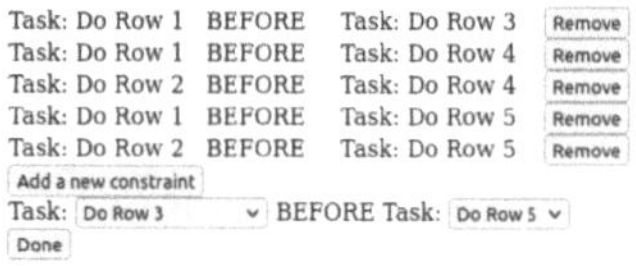

Fig. 5. Constraints definition.

bling the cell. Then, two different simple tasks are added at the lower level in the hierarchy, assuming which operator will perform them. Finally, the last step is to add information about demands through an ad hoc interface.

6   Conclusions and Future Works

We propose the extension of a knowledge engineering tool to support production engineers with a visual knowledge construction procedure to define shop-floor and HRC information. The tool is now based on SOHO and OPIS concepts to increase its effectiveness. TENANT can be useful to support also other func-tionalities to facilitate the deployment of planning and execution technologies in, e.g., a ROS environment [10]. Among future works, we aim to further inves-tigate how to leverage TENANT to investigate the dimensions mentioned in [4] and improve its effectiveness in supporting HRC applications. Also, it is essential to evaluate its usability with domain experts.

Acknowledgements. This work was supported by the Italian Ministry of Research, under the complementary actions to the NRRP "Fit4MedRob - Fit for Medical Robotics" Grant PNC0000007, (CUP: B53C22006990001).

References

1. van der Elst, S.W.G., van Tooren, M.J.L.: Application of a knowledge engineering process to support engineering design application development. In: Curran, R., Chou, S.-Y., Trappey, A. (eds.) Collaborative Product and Service Life Cycle Management for a Sustainable World, pp. 417–431. Springer London, London (2008). https://doi.org/10.1007/978-1-84800-972-1_39
2. Faroni, M., et al.: A layered control approach to human-aware task and motion planning for human-robot collaboration. In: 29th IEEE International Conference on Robot and Human Interactive Communication (RO-MAN) (2020)
3. Foderaro, E., Cesta, A., Umbrico, A., Orlandini, A.: Simplifying the a.i. planning modeling for human-robot collaboration. In: 30th IEEE International Conference on Robot Human Interactive Communication (RO-MAN) (2021)
4. Gervasi, R., Mastrogiacomo, L., Franceschini, F.: A conceptual framework to evaluate human-robot collaboration. Int. Adv. Manufact. Technol. **108** (05 2020)
5. Ha, S.H.: Applying knowledge engineering techniques to customer analysis in the service industry. Adv. Eng. Inform. **21**(3) (2007)
6. Helms, E., Schraft, R.D., Hagele, M.: rob@work: Robot assistant in industrial environments. In: Proceedings. 11th IEEE International Workshop on Robot and Human Interactive Communication (2002)
7. Jakubowski, J., Peterka, J.: Design for manufacturability in virtual environment using knowledge engineering. Manage. Product. Eng. Rev. **5**(1), 3–10 (2014). https://doi.org/10.2478/mper-2014-0001
8. Kavvathas, K., Kampourakis, E., Andronas, D., Fousekis, N., Makris, S.: A service-oriented orchestration and planning tool for plug and produce manufacturing: a deformable object handling supervision paradigm. Int. J. Comput. Integr. Manufact. **37**(12), 1626–1649 (2024). https://doi.org/10.1080/0951192X.2024.2331533
9. Ko, H., Witherell, P., Ndiaye, N.Y., Lu, Y.: Machine learning based continuous knowledge engineering for additive manufacturing. In: IEEE 15th International Conference on Automation Science and Engineering (CASE) (2019)
10. La Viola, C., Orlandini, A., Umbrico, A., Cesta, A.: Ros-tiplex: How to make experts in a.i. planning and robotics talk together and be happy. In: 28th IEEE International Conference on Robot and Human Interactive Communication (RO-MAN) (2019)
11. Niki Kousi, Spyridon Koukas, G.M., Makris, S.: Scheduling of smart intra - factory material supply operations using mobile robots. Int. J. Product. Res. **57**(3), 801–814 (2019)
12. Panetto, H., Whitman, L.: Knowledge engineering for enterprise integration, interoperability and networking: theory and applications. Data Knowl. Eng. **105**, 1–4 (2016)
13. Petrick, R.P.A., Foster, M.E.: Knowledge engineering and planning for social human–robot interaction: a case study. In: Vallati, M., Kitchin, D. (eds.) Knowledge Engineering Tools and Techniques for AI Planning, pp. 261–277. Springer International Publishing, Cham (2020). https://doi.org/10.1007/978-3-030-38561-3_14

14. Smith, S.F.: The opis framework for modeling manufacturing systems (Jun 2018). https://doi.org/10.1184/R1/6561254.v1
15. Umbrico, A., Orlandini, A., Cesta, A.: An Ontology for Human-Robot Collaboration. Proc. CIRP **93**, 1097–1102 (2020). https://doi.org/10.1016/j.procir.2020.04.045

A Tool for Generating and Labelling Domain Randomised Synthetic Images for Object Recognition in Manufacturing

Giovanna Martínez-Arellano$^{(\boxtimes)}$ and Michael G. Buck

Institute for Advanced Manufacturing, University of Nottingham, Nottingham, UK
giovanna.martinezarellano@nottingham.ac.uk

Abstract. Reconfigurable manufacturing systems are becoming the only viable option to respond to changing product volumes and product specification, which are currently major challenges for the manufacturing industry. Part of this adaptation requires vision systems to be quickly updated to handle new unseen products. For deep learning-based vision systems, this means re-training on images that might not be available. Although there is some existing work on synthetic image generation in manufacturing contexts using a variety of domain randomisation techniques, there is a lack of understanding of which domains are critical in the effectiveness of the resulting trained model. There are currently no open tools to systematically conduct such ablation studies. This paper presents a tool based on Blender and CAD models to enable the study of domain randomisation in the generation of synthetic-only datasets that can yield accurate object recognition models. Preliminary results to validate the implemented domain randomisation techniques and the ability to generate the synthetic images are presented. Once generated, synthetic data sets are used to train a YOLOv8 model for object detection as a second tool validation step. Future work will look at performing ablation studies and expanding the range of domain randomisation methods to further study the capabilities of synthetic images.

Keywords: synthetic data · cad model · domain randomisation

1 Introduction

Reconfigurable manufacturing systems (RMS) are starting to get more attention as a viable option to improve responsiveness and resilience of current manufacturing systems [1]. With the current advancements of object detection and segmentation using Machine Learning (ML) [2], it is possible for these systems to flexibly perform different tasks such as pose estimation for object pick and place [3], quality inspection [4], among others. To train such ML models, a large number of labelled images need to be available, with enough variability (noise, background, rotation, obstruction) to achieve generalisation. With a changing manufacturing environment and new product specifications, it is challenging to

© The Author(s) 2025
K. Alexopoulos et al. (Eds.): ESAIM 2024, LNME, pp. 117–124, 2025.
https://doi.org/10.1007/978-3-031-86489-6_13

have large amounts of real labelled images available. Although pre-trained models can be leveraged [5], some real images of the new object are needed. One way to address this challenge is to use synthetic images. Recent works have shown that domain-randomised, synthetic training images can yield object detection accuracy equivalent to real training images. However, there is a lack of understanding of which domains are critical for generating a fully synthetic data set that can yield such results. To understand this, more exhaustive and systematic ablation studies need to be performed across multiple manufacturing scenarios. There is currently no open implementation tool that can support such studies. Domain randomisation methods can be implemented in various ways and so open implementations need to be available to ensure comparability of ablation studies. With this context, this paper presents an open tool for the automatic generation of synthetic images and conduction of ablation studies. The tool is developed in a modular way to easily incorporate additional domain randomisation methods. To validate the implementation of the tool, tests of image generation and initial training of an object detection model using YOLOv8 have been performed. The rest of the paper is organised as follows. Section 2 provides a state-of the art on current developments in domain randomisation for synthetic images. Section 3 introduces the methodology proposed for implementing different domain randomisation techniques using Blender to manipulate the virtual scene. Preliminary results on the implementation validation of the tool are presented in Sect. 4 and Sect. 5 presents conclusions and future work.

2   Related Work

Automated pipelines for developing manufacturing-relevant synthetic images is an area of research that has got recent attention. An emerging way to generate synthetic data is through the use of generative models. Jain et al., for example, use Generative Adversarial Models to generate new images from existing real images of hot-rolled steel strips for surface defect detection [6]. Another way is through computer-aided design (CAD) models. Synthetic object data for nearly all manufactured parts is available in the form of a CAD file. Although this opens the door to automatic synthetic image generation, there is a reality gap; a model trained using synthetic images only will learn to recognise the synthetic object and not its real-life counterpart. One way to overcome this is to make the synthetic images as realistic as possible using object textures, colours, and scene lighting that matched the real object and scene [7]. Alexopoulos et al. present an automated pipeline for synthetic data generation using digital twins [8], introducing details of the real manufacturing environment to make the resulting images more realistic from a context/background point of view. These approaches assume some of the working environment is known, which may still be able to yield general deep learning models provided such models do not pick up on features of the environment itself. A contrary approach is using synthetic images that utilise the full range of visual variation that can be achieved synthetically, this is referred as domain randomisation.

Conceptually, a model trained under such high synthetic variation will see the real world as just another one of these environments [9]. Dekhtiar et al. propose a methodology based on the use of CAD models and several domain variations such as random rotation, background, saturation, contrast, brightness and blurring [10]. The authors successfully leveraged pre-trained models to classify objects from the synthetic images. The authors implement some of the known domain randomisation techniques. These works, however, do not focus on the study of the randomisation techniques and their influence in object detection accuracy. Manettas et al. propose a synthetic image generation pipeline focusing on only top and bottom views of the object and focusing on varying the rotation achieving very good accuracy from only those view points [11]. The studies presented by Tobin et al., Hinsterstoisser et al. and Trembaly et al. [9, 12, 13], present a more in depth study on the influence of different domain randomisation techniques in object detection accuracy. Through a series of ablation experiments, all these three studies conclude that the resultant models outperform their real image-trained counterparts. In each methodology, one or more of the domain randomisation parameters are excluded/weighted differently for each training data set and results are compared. Hinterstoisser et al. found that blurring and light colour are the most influential factors in detection accuracy. By varying the weight of randomisation types, Toby et al. found that the object detection accuracy was reliant on all domains except for noise. Finally, Tremblay et al. excluded randomisation types one at a time and found lighting position and textures to have the greatest effect on object detection accuracy. Overall, the three studies combined do not agree on a clear answer to the importance of each domain randomisation type. It is consistent, however, that lighting randomisation has a substantial effect on accuracy relative to other domains.

3   Methodology

In this work, a methodology for generating and labelling synthetic images using CAD models is introduced. As shown in Fig. 1, there are for image generation steps, followed by a ML model development step to enables the ablation study. The automation pipeline was developed in the 3D physics simulator Blender [14]. Within Blender, a CAD part and a virtual camera can be manipulated in a 3D environment to capture images (Step 1). Here, object, background, and lighting are fully customisable, making this software a good option for implementing modularised domain randomisation methods. This is important for allowing the user full customisation of each domain during ablation studies (source available in [15]).

3.1   Camera Positioning

For positioning the camera, the various angles from which to take the images must be decided. One potential approach is to use the 12 vertices of an icosahedron as the camera positions [12]. More vertices can be created by repeatedly

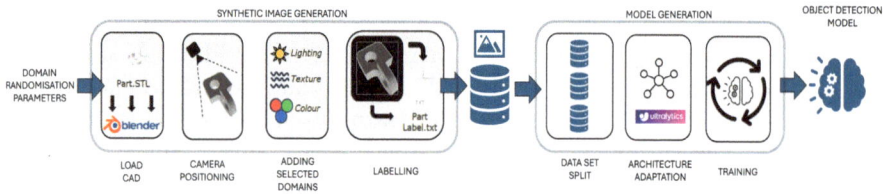

Fig. 1. Synthetic image generation pipeline used for the proposed tool using domain randomisation (DR) techniques.

splitting each face of the icosahedron to create additional vertices [16]. Although this method provides equal coverage of angles, it limits the number of angles. A more randomised approach is used by Tobin et al. [9]. Here, the position and orientation of the camera are randomised within the 3D space for each image. However, images are only captured from above the objects. In this study, a method similar to the one proposed by Dekhtiar et al. is implemented [10]. Here, the camera moves around the object on a primary axis. Each time this axis intercepts a rotation axis, the camera then follows the rotation axis, taking a set number of images of the part along the way. By allowing the user to define the number of rotation axes and image points, a unique angle can be used for every image regardless of the number of images. In addition, this approach, compared to having a fixed camera and a rotating object, avoid restraining the variation in the lighting between images.

3.2 Lighting

With the camera in position, the remaining domain randomisation features need to be set before the image is captured (Fig. 1, phase 3). Lighting is the most influential domain randomisation type for object detection accuracy according to the literature. Tobin et al. varies three light domains: number of lights, colour and position [9]. The same study also restricts the lighting conditions to those that are offered by the lights within the software. Lighting in the real world is far more complex than that generated by spotlights. In this study, a more diverse lighting method was used: high dynamic range images (HDRIs). HDRIs are a type of 360-degree image containing complex lighting [17] that can be wrapped around the scene in Blender. When used as a background, they impart the full range of lighting conditions from that image onto the object. For the tool, a random HDRI background is automatically loaded into the scene for every image taken. Each HDRI imparts lighting of varied position, colour, and intensity on the object, having complex bright and shaded areas. Thus, all three lighting domain types are randomised using one HDRI loading function. With this, and all remaining domain randomisation types added (i.e. random colour, background, texture, position and distance), the image is captured and ready for labelling.

3.3   Labelling

Accompanying all images used for training object detection models must be an image label (Fig. 1, phase 4). In this study, image label data was retrieved using the Python library OpenCV. To avoid background interference, a duplicate image is taken with the object in an identical position, but against a black background. This duplicate image is then processed using OpenCV to extract labelling information. After phase 2 (Fig. 1), the position of the object in the camera frame is unchanged. Hence, the label extracted from the duplicate image can be directly used as the label for the final image.

4   Tool Validation and Verification

To validate and verify the implemented domain randomisation techniques and the tool as whole, a test to generate a set of synthetic images given a CAD model of a real part was performed followed by the development of a deep learning model for detecting the part in a real environment. The testing of the image generation process starts with inputting first the parameters for each of the domains to be randomised. This is done via a simple Python-built user interface, where domains to randomise in the images can be selected by entering either 1 (select) or 0 (deselect), and introducing the number of images to generate. Finally, a set of background images, HDRI lighting images, and part(s) to be used by the tool are uploaded to the working directory. Once this is set up, images are generated by the tool. The following parameters were selected for verifying the correctness of the domain methods implementation:

- The number of axes for positioning the camera was 45, with 45 images being taken on each axes (45*45 = 2025 images)
- Distance was randomly selected using a set of 4 different camera focal length values. These range from the focal length where the object fills the image, to one 40 mm less (−10 mm, −20 mm, −30 mm, −40 mm).
- The object roughness, how reflective it is, and colour were randomised.
- Backgrounds that depict different examples of tables available on this github project [18] were used and selected randomly for each generated image. It is worth noting that there is no intention to replicate the real environment in the synthetic data.
- The lighting is randomly selected from a set of 3 HDRI images: a studio lit room, an indoor lit house, and a lit town at night.
- The object position is randomised in both X and Y directions.

Once generated, the images were visually inspected to verify each domain was rendering the expected results according to their implementation. Figure 2 shows some examples of the generated images. As it can be observed, changes in rotation, texture, distance are present. Texture does not seems to be particularly noticeable, but reflective properties of the object can be observed.

After this, the images were used to train a YOLOv8 model (Ultralytics Python Library). This model was chosen for being a widely used and efficient

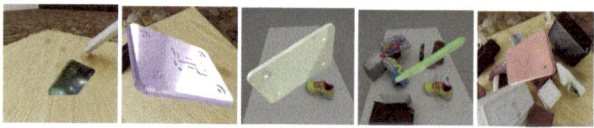

Fig. 2. Examples of images generated randomising background, lighting, position, distance and texture.

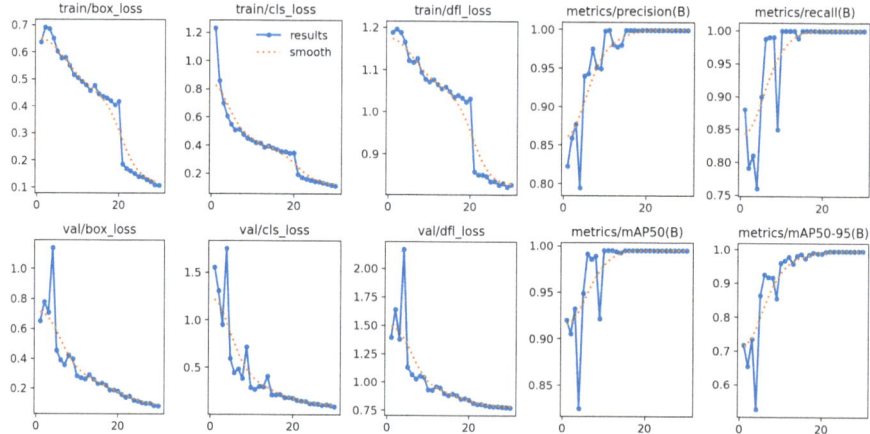

Fig. 3. Box, classification and distribution focal loss and precision during training on the training (top) and validation (bottom) sets. Validation shows mean average precision at different intersection over union thresholds.

model when working in real time [19]. The complete set of 2025 images were used for training and further 100 images were generated and used as a validation set. For testing, 20 real images taken within a robotics cell at University of Nottingham Robotics Lab where used, which contain white and black 3D printed parts corresponding to the CAD model. For using YOLOv8, the last layer was modified to introduce the class "Sensor Lid". The pre-trained model was then trained for 30 epochs as it was observed on the training/validation curves (Fig. 3) that accuracy results in the validation set were already reaching the highest precision. The resulting model was tested on the real images (some of them shown in Fig. 4). The model was able to detect 47% of the white lids but failed to detect any of the black lids. Although the colour is highly varied in the training images and is not expected to play a factor in detection, it is evident that the model struggles with this particular colour. This may be related to the YOLOv8 model itself using the colour and contrast of the object. Also, some particular angles seem to be difficult to detect. It is worth noting that no particular pose strategy was used in this initial test, which according to Hintertoisser et al. can highly increase the accuracy of the model. Despite the low accuracy, it was possible to successfully validate the implementation which then will allow thorough ablation studies to be carried out. The results highlight why it is indeed important to understand the importance of domains and suggest that some domains are more useful to be randomised and others to be more strategically used.

Fig. 4. Examples of tested real images with their corresponding bounding boxes.

5   Conclusions and Future Work

Advancing the object detection accuracy that can be achieved by using domain randomisation is the next step in facilitating object detection, and enhanced flexibility in manufacturing. In this work, a tool for testing domain randomisation for the creation of synthetic images in manufacturing is presented. The lack of consistency between the domains randomised in recent studies, and the limited industrial testing of synthetically trained deep learning models demand for a novel range of domain randomisation types, combining all of those previously tested. Preliminary results have provided an initial validation of the implementation. Future work will look at performing different ablation studies as well as to implement distractor objects.

References

1. Brunoe, T.D., Soerensen, D.G.H., Nielsen, K.: Modular design method for reconfigurable manufacturing systems. Proc. CIRP **104**, 1275–1279 (2021)
2. Zhao, Z.Q., Zheng, P., Xu, S.T., Wu, X.: Object detection with deep learning: a review. IEEE Trans. Neural Netw. Learn. Syst. **30**(11), 3212–3232 (2019)
3. Rennie, C., Shome, R., Bekris, K.E., De Souza, A.F.: A dataset for improved rgbd-based object detection and pose estimation for warehouse pick-and-place. IEEE Robot. Autom. Lett. **1**(2), 1179–1185 (2016)
4. Zheng, X., Zheng, S., Kong, Y., Chen, J.: Recent advances in surface defect inspection of industrial products using deep learning techniques. Int. J. Adv. Manufact. Technol. **113**, 35–58 (2021)
5. Lu, J., Behbood, V., Hao, P., Zuo, H., Xue, S., Zhang, G.: Transfer learning using computational intelligence: a survey. Knowl.-Based Syst. **80**, 14–23 (2015)
6. Jain, S., Seth, G., Paruthi, A., et al.: Synthetic data augmentation for surface defect detection and classification using deep learning. J. Intell. Manuf. **33**, 1007–1020 (2022)
7. Sampaio, I.G.B., Viterbo, J., Guerin, J.: Improving robustness of industrial object detection by automatic generation of synthetic images from CAD models. Comput. Intell., Article **39**(3), 415–432 (2023)

8. Alexopoulos, K., Nikolakis, N., Chryssolouris, G.: Digital twin-driven supervised machine learning for the development of artificial intelligence applications in manufacturing. Int. J. Comput. Integr. Manuf. **33**(5), 429–439 (2020)

9. Tobin, J., Fong, R., Ray, A., Schneider, J., Zaremba, W., Abbeel, P.: Domain randomization for transferring deep neural networks from simulation to the real world. In 2017 IEEE/RSJ International Conference on Intelligent Robots and Systems (IROS) pp. 23-30, IEEE, (2017)

10. Dekhtiar, J., Durupt, A., Bricogne, M., Eynard, B., Rowson, H., Kiritsis, D.: Deep learning for big data applications in CAD and PLM-Research review, opportunities and case study. Comput. Ind. **100**, 227–243 (2018)

11. Manettas, C., Nikolakis, N., Alexopoulos, K.: Synthetic datasets for Deep Learning in computer-vision assisted tasks in manufacturing. Proc. CIRP **103**, 237–242 (2021)

12. Hinterstoisser, S., Pauly, O., Heibel, H., Martina, M., Bokeloh, M.: An annotation saved is an annotation earned: Using fully synthetic training for object detection. In Proceedings of the IEEE/CVF International Conference on Computer Vision Workshops, pp. 0-0, (2019)

13. Tremblay, J., et al.: Training deep networks with synthetic data: bridging the reality gap by domain randomization. In: Proceedings of the IEEE Conference on Computer Vision and Pattern Recognition Workshops, pp. 969-977, (2018)

14. Blender: The Freedom to Create. https://www.blender.org/about/. Accessed 30 May 2024

15. Synthetic Image Generation for YOLO. https://github.com/michaelgbuck/Synthetic-Image-Generation-for-YOLO/blob/main/README.md. Accessed 30 May 2024

16. Hinterstoisser, S., Lepetit, V., Wohlhart, P.,Konolige, K.: On pre-trained image features and synthetic images for deep learning. In: Proceedings of the European Conference on Computer Vision (ECCV) Workshops, pp. 0-0, (2018)

17. Reinhard, E., Ward, G., Pattanaik, S., Debevec, P.: High Dynamic Range Imaging: Acquisition, Display, and Image-Based Lighting (The Morgan Kaufmann Series in Computer Graphics). Morgan Kaufmann Publishers Inc., S.F., CA, USA (2004)

18. Syn Table. https://github.com/ngzhili/SynTable/. Accessed 30 May 2024

19. Hussain, M.: YOLO-v1 to YOLO-v8, the rise of YOLO and its complementary nature toward digital manufacturing and industrial defect detection. Machines **11**(7), 677 (2023)

Human-Robot Interaction Through Egocentric Hand Gesture Recognition

Snehal Walunj[1]([✉]), Nazanin Mashhaditafreshi[2], Parsha Pahlevannejad[1],
Achim Wagner[1], and Martin Ruskowski[2]

[1] German Research Center for Artificial Intelligence (DFKI),
Kaiserslautern, Germany
snehal.walunj@dfki.de
[2] Rheinland-Pfälzische Technische Universität (RPTU) Kaiserslautern-Landau,
Kaiserslautern, Germany

Abstract. Recognition of human hand gestures in industrial environments is gaining popularity, especially in the context of assistance systems, thanks to advancements in deep learning-based vision methods. Also, head-worn devices with cameras are becoming more popular especially for smart assistance using Extended Reality (XR) technology, even for industrial use cases. Employing sensors from head-worn devices such as HoloLens enhance the communication between human and robot hereby providing interaction using ego-centric vision. This study delves into human-robot interaction by investigating ego-centered hand gesture recognition for commanding robots. A pipeline is developed for collecting these HoloLens video frames and to detect hand landmark labels on them using MediaPipe library by Google. Then, a Long Short-Term Memory Network (LSTM) model for hand-gesture recognition was developed that classifies the hand-gesture from the given hand landmarks in near real-time, which can then be translated into robot commands. We also present results for our network's performance and implementation pipeline.

Keywords: Egocentric Gesture Recognition · Hand Gesture Recognition · Human-Robot Interaction

1 Introduction

Hand gestures are a natural, intuitive and non-verbal communication method that humans use. These human-gestures can be translated into related robot commands [14] and enable hand gesture-based direct interface for human-robot interaction (HRI). Gesture-based control is a type of HRI system that allows a human worker to control the robot's movements using gestures in a factory environment. Vision-based recognition systems enable workers to command robots which offers exciting possibilities for collaboration between human workers and machines [15]. Extended reality (XR) in the context of worker assistance

K. Alexopoulos et al. (Eds.): ESAIM 2024, LNME, pp. 125–133, 2025.
https://doi.org/10.1007/978-3-031-86489-6_14

system on account of its ability to augment information on to real-world to support human workers. These XR based assistance systems can be coupled with a robotic assistance system [6,11], which can support workers in a smart-manufacturing environment. The head-mounted devices such as HoloLens2[1] consist of a camera that provides the egocentric view [10,13] of the person using it. This data could potentially be used for robot interaction. Since it is a head-mounted camera, it is capable of providing constant data from the moving human, unlike a fixed camera. Also unlike robot-mounted cameras where the human must necessary be in the field of view, robots and humans can collaborate and cooperate at various scenarios in smart factories. However, this research aims to leverage both static and dynamic hand gesture including single as well as double-handed gestures to interact with robots in smart factories using First-Person-View(FPV) cameras. Firstly, we need to collect data. Second, after collecting the required data, a deep learning model to classify the hand gestures should be developed. So, in the end, a pipeline can be established that includes data collection module, hand-gesture recognition module, and robot-communication module.

2   Literature Review

Ambient or fixed cameras offer the advantage of observing humans, their full-body gestures, and their environment. However, ambient camera-based hand gesture recognition (HGR) systems are often restricted by the sensor's range, requiring users to be near and/or directly in front of the camera. On the contrary, wearable camera-based HGR systems overcome this limitation due to their portability [2,7]. Wearable cameras such as head-mounted cameras provide an egocentric view of the users that makes the users always observable. However egocentric videos come with unpredictable movements and low quality due to constant motion blurs [2]. Other than RGB sensors, different kinds of data sources such as depth information can be used to enhance the quality of recognition. In [5], authors proposed a novel architecture which combines RGB and depth modalities evaluated on MECCANO dataset [12] that contains various hand gestures to mimic industrial settings. The paper [8] introduces a mobile humanoid robot that can assist humans in public spaces by following HGR using RGBD data from a robot-mounted camera. The paper [9] focuses on development and evaluation of a multistage spatial attention-based neural network for HGR which are gaining popularity. There two type of gestures, the static ones and the dynamic ones [2], also they can be further classified as single-handed or two-handed gestures. Static gestures can be easily detected using the Mediapipe which in its first step involves detecting the presence of a hand in the image. This is typically done using a CNN-based detector that identifies bounding boxes around hands. Once a hand is detected, another CNN is used to localize and predict the coordinates of specific hand landmarks within the detected hand region. MediaPipe's hand landmark model predicts 21 key points representing the knuck-

---

[1] https://www.microsoft.com/de-de/hololens.

les and fingertips.[2] library for hand pose tracking in real-time [16]. Mediapipe is used with a supporting algorithm based on Support Vector Machine (SVM)for hand-gesture recognition in [4]. In dynamic gestures the hand poses vary with time, for which Recurrent Neural Networks (RNN) are used which considers the relationship between consecutive frames. Long short Term Memory Networks (LSTMs) are type of RNNs that capture spectral, spatial as well as temporal features in a dataset [3]. Thus, it would be interesting to explore LSTMs together with CNN-based models for recognising static as well as dynamic hand gestures using single or both hands for real-time interaction with a robot, especially in a smart factory setting.

3   Problem Definition and Conceptual Approach

In smart manufacturing, a seamless interaction between human workers and robots is critical for optimizing workflow efficiency and flexibility. Gesture recognition offers a promising approach for enhancing this interaction by enabling intuitive and natural communication. In this paper, we explore the concept of using an egocentric view of the human worker-as a basis for recognizing gestures that facilitate human-robot interaction. The egocentric vision is particularly advantageous in scenarios where the human worker is physically distant from the robot, making direct visual contact or remote human-robot interaction (HRI) challenging. This capability is crucial for maintaining operational efficiency and safety in dynamic and flexible manufacturing settings, where workers may need to interact with robots from varying distances and locations. The objective of this research is to develop a comprehensive workflow for human-robot interaction using a combination of data collection, deep-learning model training, and evaluation. A method for recognizing and classifying hand gestures from an egocentric perspective is developed, utilizing a Long Short-Term Memory (LSTM) model and Mediapipe hand landmak feature extractor, with the Microsoft HoloLens2 device camera. The project also emphasizes the importance of data pre-processing, and evaluation on both recorded and real-time data. Alongside this work, an effort is made to implement gesture-based interaction between the factory worker and the robot using ROS communication.

Three classes of gestures needed to be classified. The "stop" and "come" gestures are both-handed, whereas the "continue" gesture could be done with either of the hands. The intention was to have diverse gestures. For which 1000 short videos of 39 frames for each class are recorded for the training dataset. For single handed gesture, around 500 samples were collected using the right hand, and the rest were collected using the left hand. For collecting data, Microsoft HoloLens2 is used. To save video data for further processing on a computer using HoloLens2 Sensor Streaming [1] application is utilized which transmits sensor data via TCP.

During data collection phase, different randomization techniques were included such as variable hand poses. Lighting conditions were also varied, as shown in Fig. 1. The RGB data was not relevant in dataset collection phase, since

---

[2] https://ai.google.dev/edge/mediapipe/solutions/guide

the RGB data is used by the Mediapipe to give hand landmarks. Moreover, during certain data collection sessions, intentional hand tilts were introduced to simulate real-world scenarios and data with motion blur was also introduced, as depicted in Fig. 1. Hand overlapping or occlusions is common, so various hand overlapping conditions were included. Moreover, the dataset was collected from 8 different individuals with varying hand sizes and gesture styles. These efforts aim to enhance future model training and improve the model's robustness and generalization.

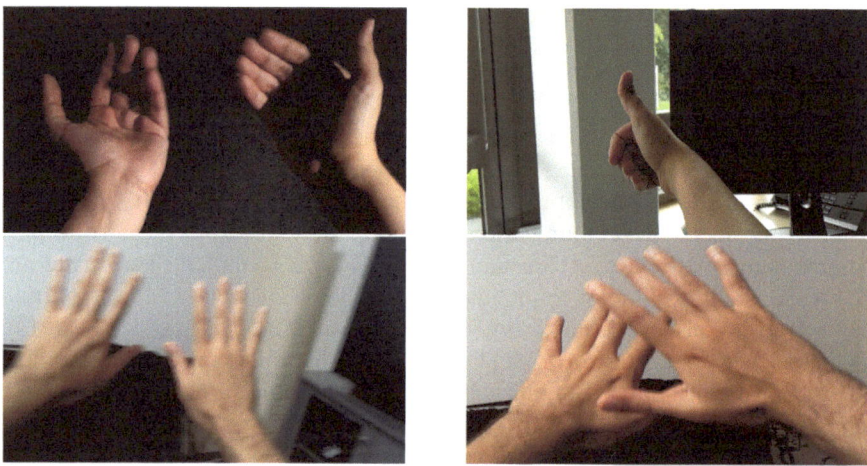

Fig. 1. Recorded data samples with different conditions of illumination, backgrounds, angles of hand poses and occlusion.

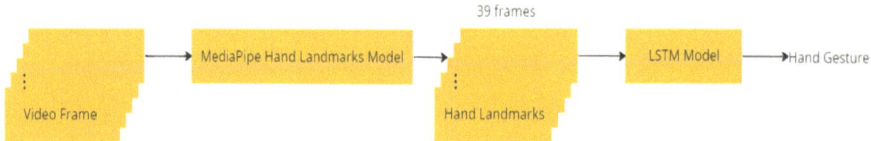

Fig. 2. Hand-gesture recognition pipeline

MediaPipe[3] already propose a CNN based approach trained on high quality and diverse dataset for hand landmark detection. It also performs very well and meet the real-time requirements. Based on the mentioned concerns and the fact that the MediaPipe hand-landmark detection model is powerful enough, a modular hand-gesture recognition approach, consisting of two modules, considered in

---

[3] https://github.com/google/mediapipe.

this project. First, recorded video frames are fed to the MediaPipe CNN model to extract the hand keypoints, and save the keypoints for each frame. Then, these keypoints will be fed into a LSTM model for gesture detection. Each video contains 39 Numpy arrays of hand landmarks. Thus a complete gesture recognition pipeline was chaled out as shown in Fig. 2.

3.1 Hand Landmarks

In the feature extraction phase, MediaPipe is used to extract hand landmarks. As illustrated in Fig. 3, MediaPipe applies a robust model based on CNN to determine the keypoint localization of 21 hand-knuckle coordinates inside the detected hand regions. Mediapipe model was trained using around 30K real-world images as well as synthetically created hand models with a variety of backgrounds [16]. A palm detection model and a hand landmarks detection model are included in the MediaPipe hand landmarker model bundle. The palm detection model detects hands inside the input image, while the hand landmarks recognition model recognizes specific hand landmarks on the palm detection model's cropped hand image.

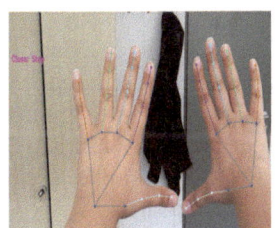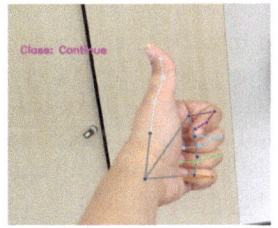

Fig. 3. From left to right, the stop, continue, and come hand gesture data labelled with hand landmarks

3.2 Classification Model

Although classification is a static problem usually tackled using CNNs alone. However humans use diverse gestures including static and dynamic ones in the most intuitive ways. In dynamic gestures such as waving, the hand pose change with time. There is noise in the input data that needs further filtering to have robust recognition of gestures which would be used as commands for the robot. Thus for the classification model a LSTM model was used together with Mediapipe feature extractor to tackle the non-linear problem of dynamic gestures. A stacked LSTM architecture was created consisting of a series of LSTM layers, followed by Dropout and Dense layers. The input to the model is a sequence of hand landmark data, represented by frames containing 21 key points, each with x and y coordinates, resulting in 42 features per frame. The model is trained to classify these sequences into one of three possible actions, as indicated by the final dense layer with a softmax activation function. SGD optimizer with learning rate = 0.001 used as an optimizer and the activation function is softmax for the last dense layer and ReLu for other units. Also Categorical Cross-Entropy was used as the loss function. The corresponding model in block diagram is shown in Fig. 4.

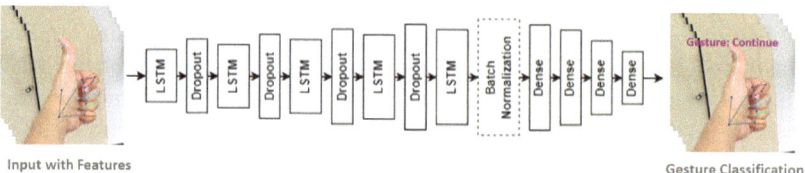

Fig. 4. Model architecture block diagram

4   Results

The final dataset was divided into 2186 samples for training, 243 for validation, and 608 for testing. The training phase is for 200 epochs, however after 10 epochs without significant improvement, training phase will stop. The model was trained in total for 82 epochs. The final test loss is 0.1287 and accuracy is 96%. In Fig. 5, the model loss and accuracy is depicted over all epochs. From the graphs, it can be derived that the training happens without overfitting or underfitting.

In Table 1 the precision, recall, and f1-score is depicted for each class on the test dataset. The performance of all classes are mostly similar with f1-score being 0.96 for 'stop', 0.98 for 'continue', and 0.95 for 'come'. The confusion matrix, as depicted in Fig. 5, summarize the performance of the classification model. Based on the confusion matrix, the model performs better on 'stop' and 'continue' in comparison to 'come'. For the 'come' gesture there was false classification as 'stop' gesture. This can me improved with more data.

5   Robot Communication

In a factory scenario where humans collaborate with robots on tasks such as maintenance, the robot is mainly used to bring the necessary tools to the human. The recognized gestures were used to command the robot. The 'come' gesture-based command is used to call the robot when the worker needs any help from the robot, provided the robot is aware of the worker's location, the 'stop' command is used to pause the robot if the help is no longer needed. The 'continue' command is used to resume the task the robot was previously performing before 'stop' command. In this way a direct human-robot interaction is carried out.

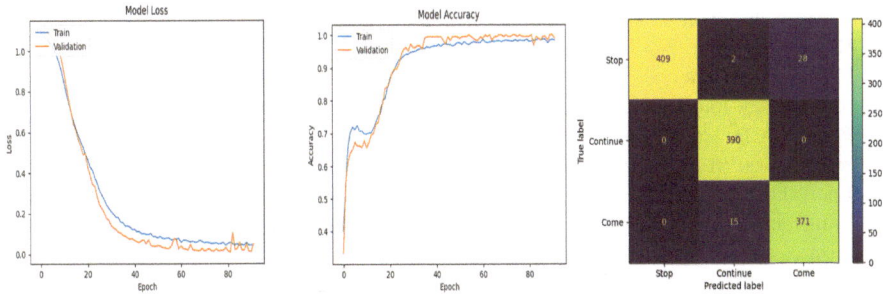

Fig. 5. Model loss and accuracy for train and validation set in 82 epochs, and confusion matrix

Table 1. Classification report on test dataset.

Label	Precision	Recall	F1-Score	Samples
Stop	1.00	0.93	0.96	439
Continue	0.96	1.00	0.98	390
Come	0.93	0.96	0.95	386

6   Conclusion and Future Work

In this research we explored the potential of ego-centric hand gesture-based human robot interaction. Google's MediaPipe library was used as the basis for hand-landmarks feature extraction. These sequences of hand landmark features were then used for a classification of dynamic hand-gestures using an LSTM neural network. The performance of the model was promising (accuracy= 0.96, loss = 0.1287) and both the total accuracy as well as classification report, for each class as an evaluation metric, were considered. Although the results seem good, the amount of false positive classification of 'come' gesture with stop gesture can be corrected using a larger dataset. Also experiments with videos with varying lengths could be used for training. More classes could be added to see how the model performs on more classes. Additionally, the potential of this gesture recognition model for hand-pose based action and activity recognition can be explored.

References

1. Dibene, J.C., Dunn, E.: Hololens 2 sensor streaming (2022)
2. Ho, H.D., Nguyen, H.Q., Nguyen, T.B., Vu, S.T., Le, T.L.: Dynamic hand gesture recognition from egocentric videos based on slowfast architecture. In: 2022 Asia-Pacific Signal and Information Processing Association Annual Summit and Conference (APSIPA ASC), pp. 01–07. IEEE (2022)
3. Hochreiter, S., Schmidhuber, J.: Long short-term memory. Neural Comput. **9**(8), 1735–1780 (1997)

4. Kavana, K., Suma, N.: Recognization of hand gestures using mediapipe hands. Int. Res. J. Modern. Eng. Technol. Sci. **4**(06) (2022)
5. Kini, J., Fleischer, S., Dave, I., Shah, M.: Egocentric rgb+ depth action recognition in industry-like settings. arXiv preprint arXiv:2309.13962 (2023)
6. Kyaw, A.H., Spencer, L., Lok, L.: Human-machine collaboration using gesture recognition in mixed reality and robotic fabrication. Archit. Intell. **3**(1), 11 (2024)
7. Le, V.D., et al..: A unified deep framework for hand pose estimation and dynamic hand action recognition from first-person rgb videos. In: 2021 International Conference on Multimedia Analysis and Pattern Recognition (MAPR), pp. 1–6. IEEE (2021)
8. Lindner, T., Wyrwał, D., Milecki, A.: An autonomous humanoid robot designed to assist a human with a gesture recognition system. Electronics **12**(12), 2652 (2023)
9. Miah, A.S.M., Hasan, M.A.M., Shin, J., Okuyama, Y., Tomioka, Y.: Multistage spatial attention-based neural network for hand gesture recognition. Computers **12**(1), 13 (2023)
10. Precup, S.A., et al.: Recognising worker intentions by assembly step prediction. In: 2023 IEEE 28th International Conference on Emerging Technologies and Factory Automation (ETFA), pp. 1–8. IEEE (2023)
11. Qian, L., Deguet, A., Wang, Z., Liu, Y.H., Kazanzides, P.: Augmented reality assisted instrument insertion and tool manipulation for the first assistant in robotic surgery. In: 2019 International Conference on Robotics and Automation (ICRA), pp. 5173–5179 (2019). https://doi.org/10.1109/ICRA.2019.8794263
12. Ragusa, F., Furnari, A., Farinella, G.M.: Meccano: a multimodal egocentric dataset for humans behavior understanding in the industrial-like domain. Computer Vision and Image Understanding (CVIU) (2023). https://doi.org/10.48550/arXiv.2209.08691, https://iplab.dmi.unict.it/MECCANO/
13. Tavakoli, H., Walunj, S., Pahlevannejad, P., Plociennik, C., Ruskowski, M.: Small object detection for near real-time egocentric perception in a manual assembly scenario. In: CVPR Workshop (2021)
14. Villani, V., Secchi, C., Lippi, M., Sabattini, L.: A general pipeline for online gesture recognition in human-robot interaction. IEEE Trans. Human-Mach. Syst. **53**(2), 315–324 (2023)
15. Walunj, S., Sintek, M., Pahlevannejad, P., Plociennik, C., Ruskowski, M.: Ontology-based digital twin framework for smart factories. In: Information Systems Development, Organizational Aspects and Societal Trends (ISD2023 Proceedings) (2023)
16. Zhang, F., et al.: Mediapipe hands: On-device real-time hand tracking (2020)

Advanced Computer Vision for Industrial Safety: Indoor Human Worker Localization Using Deep Learning

Francesco Berardinucci[(⊠)] and Marcello Urgo

Department of Mechanical Engineering, Politecnico di Milano, Milan, Italy
francesco.berardinucci@polimi.it

Abstract. Computer Vision (CV) and Machine Learning (ML) have transformed manufacturing by enabling real-time monitoring and optimization. This study introduces a novel CV-based system employing multiple RGB 2D cameras for the localization of human workers on the shop floor. The system utilizes the SCRFD pre-trained 2D person detection neural network, leveraging existing surveillance and common video cameras to monitor worker positions accurately. By tracking workers in real time, the system enhances safety by detecting hazardous situations, thereby preventing accidents. The proposed methodology was validated using videos from an industrial setting in the production of wooden house modules, demonstrating robust performance with a detection rate of 67.37% and a mean absolute error of 0.5 m. This approach provides a cost-effective and precise solution to improve worker safety and operational efficiency in manufacturing environments, advancing the integration of advanced CV techniques in industry.

Keywords: Human Worker Localization · Human Monitoring · Safety

1 Introduction

Advanced digital technologies in manufacturing environments enable the collection and analysis of vast amounts of data from sources such as sensors, cameras, and other devices in manufacturing environments. This vast amounts of data can now be harnessed to monitor and optimize manufacturing processes, improve product quality, and enhance worker safety. The integration of Artificial Intelligence (AI), Machine Learning (ML), and Deep Learning (DL) techniques has further enabled the analysis and interpretation of this data, offering valuable insights and supporting decision-making in manufacturing environments.

Building on these advancements, this work explores the use of Computer Vision (CV)-based techniques for Indoor Positioning Systems (IPS) [4] in manufacturing settings, as an alternative to traditional radio-based approaches and emerging 3D vision technologies like LIDARs. Specifically, CV-based person

© The Author(s) 2025
K. Alexopoulos et al. (Eds.): ESAIM 2024, LNME, pp. 134–143, 2025.
https://doi.org/10.1007/978-3-031-86489-6_15

detection supported by DL models is utilized to locate and monitor individuals on the shop floor. This approach offers a balanced trade-off between performance, accuracy, precision, and computational cost. The data generated not only improves worker safety by detecting and signaling hazardous situations but can also be leveraged for various other applications and services. By accurately locating workers in real time, this methodology aims to prevent accidents and enhance overall safety in the workplace, marking an advancement in the practical application of CV and DL technologies in manufacturing environments.

2   Related Works

Recent advancements in computer vision and deep learning have enabled real-time object detection using video cameras. OpenPose [1] was the first real-time multi-person system to detect 135 human body, hand, facial, and foot keypoints in single images, estimating human poses in real-time. Since then, many AI techniques for human pose detection and analysis have seen significant application in the industrial sector [19], particularly for safety and manufacturing process monitoring [20]. The lack of specialized datasets for Person Search [18] often leads to the use of general object detection datasets and models. Common approaches involve fine-tuning pre-trained models like YOLO [13] or Faster R-CNN [14] on custom datasets, with examples applied to manufacturing workers identification [7]. Additionally, datasets for sub-tasks such as Pedestrian Detection [9] and Re-Identification [5] have been effective.

Alternative strategies for worker detection employ Radio Frequency (RF) and Light Detection and Ranging (LIDAR) technologies. RF-based systems use radio waves to locate workers in real-time, with applications in industrial environments [11,12]. LIDAR systems use laser light to measure distances and create 3D maps of the environment, with applications in worker safety and monitoring [15]. While these technologies offer high accuracy and reliability, they are often expensive and require specialized equipment, limiting their widespread adoption in manufacturing environments.

InsightFace [3,6] offers a deep learning-based facial detector library for Python. Despite its initial design for face detection, InsightFace models have proven effective in detecting workers in real-time video streams on shop floors [10]. Combining these models achieves high accuracy and robustness in tracking people on shop floors, integrating with other safety systems like machine learning-based fall detection for comprehensive safety monitoring.

3   Methodology

The proposed methodology aims to localize indoor human workers in a manufacturing environment by employing commercially-available monocular RGB cameras. The acquired single-view videos are processed using CV techniques and analyzed using a DNN person detection model to finally derive workers location

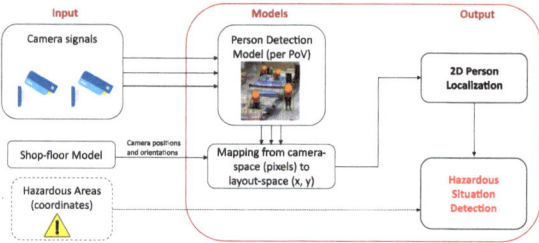

Fig. 1. Schema of camera-based positioning and hazardous situation detection.

in the 2D factory floor space. The approach supports the utilization of multiple RGB 2D cameras strategically positioned to provide redundant coverage of the same area, providing multiple detections of the same worker in different camera views, without identifying the worker. This redundancy is critical for ensuring comprehensive visibility and reliability in data acquisition, even in the presence of potential occlusions or environmental changes that are common in a manufacturing environment. The combined perspectives from these cameras create a robust framework for real-time monitoring and analysis where, if at least one detection is found in an hazardous area, an alarm can be triggered. The following sections detail the sequential steps of the approach, from preprocessing video signals to inferring multiple workers' positions on the shop-floor layout, that are crucial to develop hazard detection systems and improve worker safety.

The methodology is designed grounding on the following assumptions: (i) At least two cameras should cover the considered area; (ii) Cameras must be calibrated and synchronized.

Since the objective is the localization of workers rather than their identification, this approach is more viable from a regulatory perspective and less intrusive for workers' privacy while still ensuring necessary safety features, at the cost of possible multiple detections of the same worker.

The approach comprises the following steps: (1) Preprocessing of the video stream (distortion correction, perspective transform derivation); (2) Detection of human beings in the video stream; (3) Localization of one or more human operators on the shop-floor space. The detailed process steps are presented in the following paragraphs. Figure 1 provides an overview of the approach structure, integrated in a hazardous situation detection application. The human workers are localized using the processed video streams and DNN model, thus obtaining 2D location coordinates of the shop-floor. Externally, the system receives identified hazardous regions' coordinates in the same reference system. Safety hazards are detected by testing workers' positions in the defined regions.

3.1 Preprocessing of Video Signals from Calibrated Cameras

The first phase consists in preprocessing the video signals of calibrated and synchronized cameras. This step involves synchronizing the video signals, correcting

Fig. 2. Video frame showing the distortion correction using the intrinsic calibration parameters of the camera lens and the resulting undistorted image.

the lens distortions and aberrations, to be able to estimate the spatial configuration of the cameras. The first requirement for measuring using cameras is to work on calibrated video streams. Thus, the geometrical distortion caused by the camera lenses must be corrected. This correction requires a suitable mathematical model for the lens to model the introduced distortion and the associated parameters. Data supporting this correction are sometimes available by the lens manufacturer, by external databases [17] or measured by performing an *intrinsic calibration* of the lens [2]. A sample distortion correction result is shown in Fig. 2.

3.2 Person Detection, DNN Occlusion Inference and Positioning

The next phase consists in the detection of persons in the video stream and the extraction of bounding boxes in pixel space. This step involves using a DL model to analyze each distortion-corrected frame of the videos and identify the presence of humans in the scene. The output of the neural network consists in a vector of $(4 + 4) * n$ keypoint values in frame space (percentages), corresponding to the vertex coordinates of two bounding box per person detected (Fig. 3). The first set contains a bounding box for each visible person in the frame (in green), while the latter are estimates of the entire person body, including possibly occluded body parts (in red). The second set will be used for the localization to ensure robustness against occlusions. Each bounding box is defined by its four vertex points in pixel coordinates. For this detection, the InsightFace SCRFD neural network model [6] has been selected after validating its performances in videos with one or multiple operators present and with partially occluded bodies. The model uses a ResNet backbone and accepts RGB images as input (w × h × 3). It was selected after testing performances of multiple models in videos where one or multiple operators where present and with partially occluded bodies and after the performance evaluation that showed that the model is capable to be run in real-time (see Sect. 4) which is a requirement for safety applications, and the SCRFD network's capability to estimate the entire body bounding box, even in the presence of occlusions, in addition to the visible body bounding box.

The objective is to obtain operators' location in real-word coordinates, therefore it is wanted to derive a point value from the detections. To achieve this, the worker's feet location will be used as an approximation for his/her position on the factory floor. This point value (in pixel-space) is derived, specifically, by considering the middle point of the lower side of the bounding box of the whole

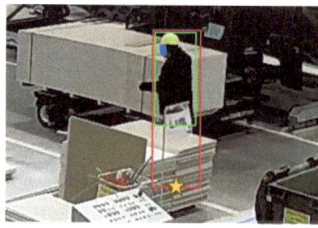

Fig. 3. Person detection model output with partial occlusion. The considered position point is shown (star), based on the inferred body bounding box.

human body (Fig. 3, star symbol), assuming that the worker is operating on the factory floor. Additionally, using the estimated bounding box provides robustness against occlusions of the lower part of the body.

Considering this point as the position of the operator in the camera pixel-space, his/her successive locations in the shop floor are derived by projecting his/her coordinates in the shop-floor-space, using the described extrinsic projection matrix (Sect. 3.3).

3.3 Transformation of Bounding Boxes Coordinates from Pixel Space to Real-World 2D Space and Reconstruction of Shop Floor Status

The final step in the methodology is the transformation of the derived reference location point estimates (Fig. 3) from pixel space to real-world 2D space and the reconstruction of the shop floor's current status. This allows for a reconstruction of the current status of the factory floor, including the location of the detected operators. Known the camera angle of view and direction in real world coordinates, each point is projected from pixel space to the real-word space by a perspective transform.

Specifically, It is necessary to derive a mathematical function to map each point in the pixel space to the shop-floor coordinate space. For this application, since the objective is to locate the operators on the shop floor, a 2D cartesian space is considered, corresponding to the physical floor of the factory. An extrinsic calibration is required to derive the mathematical function of the projection between the two spaces. To this aim, the parameters listed in Table 1, related to the position of the camera on the shop floor, are required. Using these parameters, it is possible to project each point defined in the *pixel-space* to the *shop-floor-space*, by deriving a rotation matrix $\mathbf{R}$ and a translation vector $\mathbf{t}$. The extrinsic matrix $\mathbf{P}$ is obtained by appending the vector $\mathbf{t}$ to the $\mathbf{R}$ matrix, then the extrinsic matrix is used to model the relation between the camera pixel-space and the shop-floor-space.

Table 1. Input parameters of the extrinsic calibration.

Parameter	Unit of measure
α_i: Camera angle of view	degrees
(w, h): video resolution	pixels
θ_i: Orientation of camera i (vertical axis, yaw)	degrees
ψ_i: Orientation of camera i (pitch)	degrees
P_i: 2D position of camera i (in world coordinates)	meters
CSYS: Shop-floor coordinate system	—

The extrinsic matrix is used to model the relation between the camera pixel-space and the shop-floor-space:

$$\mathbf{P} = [\mathbf{R}|\mathbf{t}] \qquad \begin{bmatrix} X_c \\ Y_c \end{bmatrix} = \mathbf{P} \begin{bmatrix} X_w \\ Y_w \\ 1 \end{bmatrix} \tag{1}$$

An example frame showing the transformation applied to each pixel of the input image is presented in Fig. 4. Finally, the detected position of the worker (the midpoint of the lower segment of the bounding box, Sect. 3.2) is projected on the shop floor frame-by frame, as shown in Fig. 5. The resulting trajectory of positions can then be employed with external systems and downstream applications to detect hazardous situations in the shop floor by comparing the worker's position with the location of dangerous areas.

4 Validation and Results

The described approach has been tested in an industrial setting involving the construction of wooden house modules. The videos come from surveillance cameras installed in the factory, above the module assembly area, directed towards a crane loading/unloading area. This area was selected due to the potential safety risks for workers handling the house module components. A control region was defined within this area, prohibiting worker access during crane operations. To test the methodology, multiple videos have been recorded in which a single

Fig. 4. Video frame illustrating the perspective transform and projection of the image on the shop-floor plane using extrinsic calibration parameters.

Fig. 5. Test output. The distortion-corrected and re-projected floor is shown. On the foreground are overlaid the detected worker positions and their coordinates.

worker is operating, visible from two cameras' fields of view. The method is tested on the complete video, a 30 s recording (30 fps) containing a total of 900 image frames.

The intrinsic calibration has been performed by selecting a set of intrinsic parameters of a suitable equivalent lens, adapted by the crop factor, from the LensFun database [17]. As the obtained calibration has resulted to be reasonably accurate for the application, the undistorted video stream has been further processed according to the described approach. Nevertheless, an on-site intrinsic calibration could further improve the accuracy of the results. The extrinsic calibration was performed using existing references available from the camera point of view (machinery, road signs) for which a CAD model is available. To implement the approach in areas where such references are not available, temporary visual markers on the shop floor are needed. The used SCRFD model implementation is InsightFace, developed using PyTorch and provided by the authors of SCRFD [6] along with the model weights trained on the CrowdHuman dataset [16].

In Fig. 5, a complete output from one of the recorded videos is presented, where the complete detected path of the worker is plotted (in blue). The results show a robust detection of the workers and a reasonable accuracy for the considered safety application. The performances meet the requirements for running sustainably in real-time, since the model can run at approximately 60–80 fps per single core on an Intel i9 CPU, while surveillance videos are usually recorded at 15–30 fps. The quantitative results on the analyzed experimental test are reported in Table 2. Overall, the model is able to detect the walking operator in approximately two-thirds of the total frames. If continuous trajectory data is required, an interpolation and/or filtering must be performed. The localization achieves a mean error of approximately 0.5 m. For reference, in Fig. 6a the detected trajectory of the operator on the shop-floor is reported, alongside the ground truth.

Additionally, the absolute errors per frame, computed as the norm of the distance vector between the real and detected positions are shown in Fig. 6b. It is visible, when comparing the diagram to the trajectory, that the absolute

Table 2. Quantitative results for the test video experiment.

Metric	Value
Detection Rate	67.37%
Mean Absolute Error [pixel]	33.39
Mean Absolute Error [mm]	534.37
Average FPS (CPU, single-threaded)	~65fps

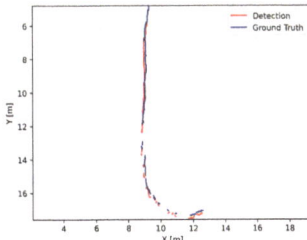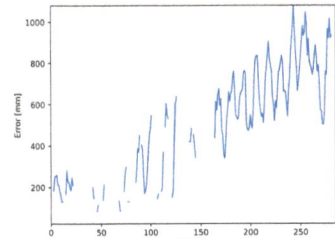

(a) Ground-truth and detected operator paths, projected on the shop-floor.

(b) Absolute errors for each frame. Values are missing for frames with no detection.

Fig. 6. Comparative analysis of the results.

errors increase with the distance of the worker with respect to the camera, as expected. Furthermore, oscillations are visible due to the strides of the operator since the position is derived from the bounding box of the body in the frame. An ad-hoc filtering could be implemented to filter out the oscillations, if necessary.

5 Conclusions

In this work several key findings are observed. The core of the vision-based solution, revolves around the utilization of video signals to accurately localize workers. This process begins with the application of machine learning techniques to detect and identify humans within the camera feed. Following the successful detection, the next step involves estimating the location of the detected personnel directly from the video frames. However, it is important to note that the detection algorithm's performance, while commendable, achieves correct detection in approximately two-thirds of cases on average. This level of performance, although significant, introduces gaps in the obtained trajectory data, underscoring the necessity for advanced trajectory estimation methods to fill these gaps, increasing the complexity of the development. Additionally, the limited field of view provided by a single camera further complicates the detection process, making it clear that a network of multiple cameras is essential to ensure continuous and comprehensive coverage across the entire area of interest. Nevertheless, thanks to the chosen per-video analysis strategy and without performing person

identification, introducing redundancy through multiple cameras do not complicate the approach, but rather enhance its robustness and reliability, especially in
safety applications.

One of the most notable advantages of this camera-based method over alternative solutions, such as those based on RF and LIDARs, is the simplicity of
implementation, by using non-specialized equipment, often already installed in
industrial environments. Additionally, the structure of the approach tailored to
work with multiple single-view localizations, allow to optionally integrate in the
system alternative positioning systems if needed. Further developments could
include data fusion with other sensors, such as RF or LIDAR, to enhance the
system's robustness and reliability. Additionally, the prediction of the worker trajectory [8] can be investigated to predict hazardous situations ahead of time and
to identify near-misses. Finally, by implementing a depth estimation algorithm,
the system could avoid the need for extrinsic calibration, further simplifying the
setup process and allow for simple reconfiguration.

Acknowledgements. This work was partially funded by the European HE research
project 5G-TIMBER, GA no. 101058505. The authors thank Harmet OÜ for their
support in defining the case study.

Declaration of Interests. The authors have no competing interests to declare that
are relevant to the content of this article.

References

1. Cao, Z., Hidalgo, G., Simon, T., Wei, S.E., Sheikh, Y.: OpenPose: realtime multi-
person 2D pose estimation using part affinity fields. arXiv:1812.08008 (2019)
2. De la Escalera, A., Armingol, J.M.: Automatic chessboard detection for intrinsic
and extrinsic camera parameter calibration. Sensors **10**(3), 2027–2044 (2010)
3. Deng, J., Guo, J., Ververas, E., Kotsia, I., Zafeiriou, S.: RetinaFace: single-shot
multi-level face localisation in the wild. In: 2020 IEEE/CVF Conference on Computer Vision and Pattern Recognition (CVPR), pp. 5202–5211. IEEE, Seattle, WA,
USA (2020)
4. Farahsari, P.S., Farahzadi, A., Rezazadeh, J., Bagheri, A.: A survey on indoor
positioning systems for IoT-based applications. IEEE Internet Things J. **9**(10),
7680–7699 (2022)
5. Fu, D., et al.: Unsupervised pre-training for person re-identification. In: 2021
IEEE/CVF Conference on Computer Vision and Pattern Recognition (CVPR),
pp. 14745–14754. IEEE, Nashville, TN, USA (2021)
6. Guo, J., Deng, J., Lattas, A., Zafeiriou, S.: Sample and computation redistribution
for efficient face detection (2021)
7. Hu, J., Gao, X., Wu, H., Gao, S.: Detection of workers without the helmets in
videos based on YOLO V3. In: 2019 12th International Congress on Image and
Signal Processing, BioMedical Engineering and Informatics (CISP-BMEI), pp. 1–4
(2019)
8. Katsampiris-Salgado, K., Dimitropoulos, N., Gkrizis, C., Michalos, G., Makris, S.:
Advancing human-robot collaboration: Predicting operator trajectories through AI
and infrared imaging. J. Manuf. Syst. **74**, 980–994 (2024)

9. Khan, A.H., Nawaz, M.S., Dengel, A.: Localized semantic feature mixers for efficient pedestrian detection in autonomous driving. In: 2023 IEEE/CVF Conference on Computer Vision and Pattern Recognition (CVPR), pp. 5476–5485. IEEE, Vancouver, BC, Canada (2023)
10. Liu, W., Liao, S., Ren, W., Hu, W., Yu, Y.: High-level semantic feature detection: a new perspective for pedestrian detection. In: 2019 IEEE/CVF Conference on Computer Vision and Pattern Recognition (CVPR), pp. 5182–5191. IEEE, Long Beach, CA, USA (2019)
11. Müürsepp, I., et al.: Performance evaluation of 5G-NR positioning accuracy using time difference of arrival method. In: 2021 IEEE International Mediterranean Conference on Communications and Networking (MeditCom), pp. 494–499 (2021)
12. Pilati, F., Sbaragli, A.: Learning human-process interaction in manual manufacturing job shops through indoor positioning systems. Comput. Ind. **151**, 103984 (2023)
13. Redmon, J., Divvala, S., Girshick, R., Farhadi, A.: You only look once: unified, real-time object detection. In: 2016 IEEE Conference on Computer Vision and Pattern Recognition (CVPR), pp. 779–788 (2016)
14. Ren, S., He, K., Girshick, R., Sun, J.: Faster R-CNN: towards real-time object detection with region proposal networks (2016)
15. Roth, M., Jargot, D., Gavrila, D.M.: Deep end-to-end 3D person detection from camera and lidar. In: 2019 IEEE Intelligent Transportation Systems Conference (ITSC), pp. 521–527. IEEE, Auckland, New Zealand (2019)
16. Shao, S., Zhao, Z., Li, B., Xiao, T., Yu, G., Zhang, X., Sun, J.: CrowdHuman: a benchmark for detecting human in a crowd (2018)
17. Tang, Z., Grompone von Gioi, R., Monasse, P., Morel, J.M.: A precision analysis of camera distortion models. IEEE Trans. Image Process. **26**(6), 2694–2704 (2017)
18. Tian, Y., Chen, D., Liu, Y., Yang, J., Zhang, S.: Divide and Conquer: hybrid pre-training for person search (2023)
19. Urgo, M., Berardinucci, F., Zheng, P., Wang, L.: AI-based pose estimation of human operators in manufacturing environments. In: Tolio, T. (ed.) CIRP Novel Topics in Production Engineering, vol. 1, pp. 3–38. Springer Nature Switzerland, Cham (2024)
20. Urgo, M., Tarabini, M., Tolio, T.: A human modelling and monitoring approach to support the execution of manufacturing operations. CIRP Ann. **68**(1), 5–8 (2019)

AI in System Level

LOS Data Set: A Large Scale Online Scheduling Benchmark for Flexible Job Shop Problems with Setup and Transportation Times

Katharina Hengel[1]([✉]), Achim Wagner[1], and Martin Ruskowski[1,2]

[1] German Research Center for Artificial Intelligence (DFKI),
Kaiserslautern, Germany
katharina.hengel@dfki.de
[2] Technologie-Initiative SmartFactory KL e.V., Kaiserslautern, Germany

Abstract. With an increased flexibility in the production new scheduling techniques are necessary to accommodate this change. Though there have already been published many scheduling algorithms fostering this demand for flexibility, there is no common ground on a benchmark data set to compare these approaches against each other. Therefore, this paper aims at the generation of a benchmark data set for the flexible job shop problem (FJSP) with setup and transportation times on which different scheduling algorithms can be evaluated. The data set is specified by several key parameters from which FJSP are created. The use and advantage of the large-scale online scheduling (LOS) data set is exemplified by its application on a Reinforcement Learning online scheduling algorithm and dispatching rules. Furthermore, backward compatibility is established with the former FJSP notation.

Keywords: Benchmark · Scheduling · Flexible Job Shop · Data Set

1 Introduction

Industry 4.0 increases the flexibility of the production in the shop floor. Hence, new approaches for scheduling are required to accommodate for the accompanying changes introduced in Industry 4.0. On the hardware side the flexibility is often enabled by the use of modular production systems. In this connection, the resulting scheduling problem is formally defined by the flexible job shop problem (FJSP) with setup and transportation times. Though there have already been published many scheduling algorithms fostering this demand for an increased flexibility, there is no common ground on a benchmark data set to compare these approaches against each other.

Towards this end, the data sets which are used are either not published and explained in detail, small scale or only have few instances. This makes a broad adaptability difficult. Hence, a comparison between the algorithms as well as

K. Alexopoulos et al. (Eds.): ESAIM 2024, LNME, pp. 147–156, 2025.
https://doi.org/10.1007/978-3-031-86489-6_16

the evaluation of its advantages and disadvantages is not possible. Therefore, we present a benchmark data set for the FJSP with setup and transportation times on which different scheduling algorithms can be evaluated.

The data set is specified by several key parameters from which FJSPs are created. A new notation is introduced for better extensibility and readability reasons. The use and advantage of the large-scale online scheduling (LOS) data set is exemplified by its application on a Reinforcement Learning online scheduling algorithm and dispatching rules. Furthermore, backward compatibility is established with the former FJSP notation. As example the existing FJSP data set published by [12] is converted to our notation. This allows the comparison of the results between previous and newly published algorithms using either of these notations.

The remainder of this paper is organized as follows. An introduction to the FJSP with setup and transportation times as well as an overview for existing scheduling data sets is given in Sect. 2. Afterwards the method for the creation of the FJSP data set is introduced in Sect. 3. At last a RL scheduling algorithm and dispatching rules are applied on the data sets. The results are demonstrated in Sect. 4 and discussed in Sect. 5.

2   State-of-the-Art

2.1   Flexible Job Shop Problem

An overview of the taxonomy of scheduling problems is given in [14]. To handle a high degree of flexibility in the shop floor we focus on online scheduling for FJSP with setup and transportation times.

Given are n jobs $J = \{j_1, j_2, ..., j_n\}$. Each job j_i is composed of r operations $O^i = \{o_1^i, o_2^i, ..., o_r^i\}$. Each operation o_u^i must be processed before every operation o_v^i with $u < v$. Every operation must be processed on one of the machines $M = \{m_1, m_2, ..., m_h\}$. A machine can only process one operation at a time. While an operation o_u^i in the general job shop problem (JSP) must be processed on one specific machine, the FJSP relaxes this condition and an operation can be process by a specified subset $M^{iu} \subset M$ of all machines. The processing time of an operation o_u^i on machine l is given by p_l^{iu}. Between the processing of two different jobs on a machine, a setup time might occur. The setup time is denoted by $c_l^{iu,vw}$ with l being the machine id and o_u^i and o_w^v being the operations. Furthermore, t_{lk} denotes the transportation time of a job between the machines l and k.

2.2   Benchmarks

Figure 1 show an overview of publicly available data sets applicable for testing FJSP scheduling algorithms. Since the FJSP is a generalization of the JSP, every flexible job shop scheduling algorithm can also be applied to a JSP instance. A large data set of JSPs is contained in the OR-library [4]. It includes JSP by [1,2,8,10,11,15,16]. Since the OR-library's JSP instance are rather small, [6]

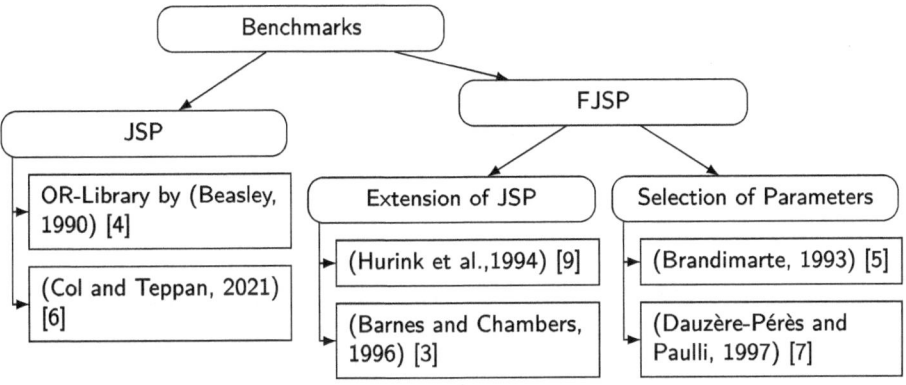

Fig. 1. Taxonomie of benchmark data sets

created a large scale set of JSP. In their first approach they hereby extended the benchmark set of [16].

The FJSP instances can be differentiated into two groups:

1. Existing JSP instances can be extended to the FJSP case by defining M^{iu}
2. FJSP can be created from scratch by selecting parameters withing given ranges

[3,9] are examples of the first group which extend existing JSP instances from [1,8] resp. [8,10]. The JSP differs from the FJSP only in the definition of M^{iu}, i.e. the operation o^i_u can be processed by multiple machines $m \in M^{iu}$ with $|M^{iu}| \geq 1$. In contrast $|M^{iu}| = 1$ holds in the JSP case. Hence, only M^{iu} must be redefined. To do so [9] fixes $\operatorname*{avg}_{i,u}|M^{iu}|$ and $\operatorname*{max}_{i,u}|M^{iu}|$. By using the parameter selection depicted in Table 1 four different subsets are created, named sdata, edata, rdata and vdata. Hereby, the edata represents the instances where only few operations can be processed by different machines. In the rdata most operations can be assigned to a few number of different machines and in the vdata each operation can be assigned to many different machines.

Table 1. Parameter assignment for the FJSP instances by [9]

	$\operatorname*{avg}_{i,u}|M^{iu}|$	$\operatorname*{max}_{i,u}|M^{iu}|$
sdata	1	1
edata	1.15	2, if $h \leq 6$ 3, otherwise
rdata	2	3
vdata	0.5 h	0.8 h

On the other hand in [3] the cumulative processing time (CPT) and the number of critical operations are utilized to replicate machines of one JSP instance. More specifically an instance is modified in one of the following ways:

- the machine requiring the greatest CPT is replicated once
- the machine requiring the greatest CPT is replicated twice
- the machine requiring the greatest CPT is replicated three times
- the machines requiring the greatest and second-greatest CPTs are replicated once each
- the machines requiring the greatest, second-greatest, and third-greatest CPTs are replicated once each
- the machine with the greatest number of critical operations is replicated once
- the machines with the greatest and second-greatest number of critical operations are replicated once each

Table 2. Parameter selection of [5,7]

Parameter	[5]	[7]
$n = \|J\|$	$n \in \mathbb{N}$	$n \in \mathbb{N}$
$h = \|M\|$	$h \in \mathbb{N}$	$h \in \mathbb{N}$
$r = \|O^i\|$	$r \in_R [r_{min}, r_{max}]$	$r \in_R [r_{min}, r_{max}]$
	$r_{min} \in \mathbb{N}, r_{max} \in \mathbb{N}$	$r_{min} \in \mathbb{N}, r_{max} \in \mathbb{N}$
M^{iu}	$\|M^{iu}\| \in_R [1, b^{iu}_{max}], b^{iu}_{max} \in \mathbb{N}$	$P(l \in M^{iu}) \in [0,1]$
p^{iu}_l	$p^{iu}_l \in_R [p_{min}, p_{max}]$	$\overline{p^{iu}} := \frac{\sum_{l \in M^{iu}} p^{iu}_l}{\|M^{iu}\|} \in [\overline{p_{min}}, \overline{p_{max}}]$
	$p_{min} \in \mathbb{N}, p_{max} \in \mathbb{N}$	$\overline{p_{min}} \in \mathbb{N}, \overline{p_{max}} \in \mathbb{N}$
		with $\|p^{iu}_l - p^{iu}_k\| \leq \Delta p_l \forall l, k \in M^{iu}, \Delta p_l \in \mathbb{N}$

In case no JSP instance is the basis for the FJSP creation, all FJSP parameters must be set. [5,7] do so by using a set of parameters and boundaries within which they are selected. Table 2 presents a comparison of both approaches. Let $x \in_R S$ denote that x was selected from R uniformly at random while $x \in S$ denotes that x was selected manually from S. In [7] a probability $P(l \in M^{iu}) \in [0,1]$ is set for every M^{iu}. P is the probability of a machine l being in M^{iu}. In the case M^{iu} would be empty a machine $\bar{l}$ is selected randomly and $M^{iu} = \{\bar{l}\}$.

A collection of FJSP instances by [3,5,7,9] is provided by [12]. It encompassses in total 313 benchmark instances in standard FJSP notation syntax. All instances use the standard FJSP notation syntax. In the first line the number of jobs and the number of machines is specified. It is followed by a line for each job. Each job specification starts with the number of operations of the job. It is followed by a sequence for each operation specifying the number of machines which are able to process the operation and the machine ids and the processing times of them. An example of this notation is given below using the instance mt06 by [8]:

```
6 6 1
6 13 1 11 3 12 6 14 7 16 3 15 6
6 12 8 13 5 15 5 10 16 10 1 1 10 14 4
6 13 5 14 4 16 8 11 9 12 1 1 15 7
6 12 5 11 5 13 5 14 3 15 8 16 9
6 13 9 12 3 15 5 16 4 11 3 14 1
6 12 3 14 3 16 9 1 1 10 15 4 13 1
```

3   Method

A job is defined by a sequence of operations. While an operation describes "the application of a skill on a defined product type with a desired outcome" [13], the term skill defines "the ability of a resource to perform a process" [13]. Hence, the term skill relates to processes and resources while the term operation relates to products. Even trough the represent the same, they show different viewpoints on the process. By using a fixed set of skills $S = \{s_1, s_2, ..., s_g\}$ instead of operations in the definition of the FJSP, reusable building blocks can be formulated. Instead of defining the processing time for an operation o_u^i, we define the processing time of a skill s_u, i.e. p_l^u. Furthermore, we define the setup time c_l^{uv} of a machine between skills s_u, s_v and a job as sequence of skills. In this way, we can scale the number of jobs and machines upwards while limiting the skills without a blow-up of the instance specification.

For the generation of our benchmark data set we used approaches from [5,9]. We defined the parameters for the FJSP creation as follows:

- $n \in_R [n_{min}, n_{max}], n_{min} \in \mathbb{N}, n_{max} \in \mathbb{N}$
- $h \in_R [h_{min}, h_{max}], h_{min} \in \mathbb{N}, h_{max} \in \mathbb{N}$
- $g \in \mathbb{N}$
- $r \in_R [r_{min}, r_{max}], r_{min} \in \mathbb{N}, r_{max} \in \mathbb{N}$
- $\underset{i,u}{\text{avg}} |M^{iu}| \in \mathbb{R}^+$
- $\underset{i,u}{\text{max}} |M^{iu}| \in \mathbb{R}^+$
- $p_l^u \in_R [p_{min}, p_{max}], p_{min} \in \mathbb{N}, p_{max} \in \mathbb{N}$
- $c_l^{uv} \in_R [s_{min}, s_{max}], s_{min} \in \mathbb{N}, s_{max} \in \mathbb{N}$

Furthermore, let $\tau \in \mathbb{N}$ be a scaling factor for the transportation times. The machines are arranged in a square matrix inside the factory. The distance between two machines is then defined by the manhattan distance between the machines multiplied by the scaling factor for the transportation times. The skills of a machine are selected following the approach of [9]. Using a triangular distribution with lower limit 1, upper limit $\underset{i,u}{\text{max}} |M^{iu}|$ and mode $\underset{i,u}{\text{avg}} |M^{iu}|$ the number of machines per skill is defined. Afterwards the selected number of machines are sampled randomly from the set of machines. The processing time of the machine for the skill is randomly sampled from the uniform distribution between the given limits. The setup times are selected randomly using a uniform distribution within the given limits. For the definition of the jobs we select the number of jobs, for each job the number of skills and for each skill the skill itself uniformly at random.

With the utilization of skills, we could not follow the standard FJSP notation. Instead, we used a yaml based syntax for the LOS data set instances. This also allows easier extensibility and readability. Each FJSP is described by a map with the keys distances, skills, changeovers and jobs. The values are represented as (nested) lists. To ensure the comparison with benchmarks specified by the standard FJSP notation, we established backwards compatibility with it. By setting the setup and transportation times to 0 and mapping every operation in the former notation to a different skill in our yaml notation, the comparison to previous used benchmarks can be established.

4   Results

We applied the dispatching rules of Table 3 and a scheduling algorithm using Reinforcement Learning (RL) on the benchmark data set of [12] as well as on the LOS data set. Since the setup and transportation times are 0 for the instances in [12], the dispatching rules SST and SST + SPT are not applied in this case. In Figs. 2, 3, 4, 5 and 6 the resulting make span of the scheduling algorithms are presented. The results are illustrated as box plot created by 21 FJSP instances of [3], 10 of [5], 18 of [7], 264 of [9] and 100 of the LOS data set.

Table 3. Overview of dispatching rules

Abbreviation	Dispatching Rule	
SIRO	Setup in Random Order	
SST	Shortest Setup Time	$\min(c_l^{iu,vw} + t_{lk})$
SPT	Shortest Processing Time	$\min p_l^{iu}$
SPT + SST	Shortest Processing + Setup time	$\min(p_l^{iu} + c_l^{iu,vw} + t_{lk})$

For the results of the data set of [12] (Figs. 2, 3, 4 and 5) the results do not vary much between the different algorithms. The boxes as well as the whiskers overlap most of the time. The largest difference of the mean respectively the median occurs in the data set of [7]. Here the median differs between the RL algorithm and the SPT dispatching rule by 1063, the mean respectively by 1922.5. While RL performs the best in the data set of [3], SPT has the best performance in the data sets of [7,9]. In [5] RL and SPT perform nearly the same.

For the evaluation of the LOS data set, we created over 8000 FJSP instances with the following parameter assignment: $n = 100$, $h = 25$, $g = 25$, $r \in_R [5,8]$, $\text{avg}_{i,u} |M^{iu}| = 12.5$, $\max_{i,u} |M^{iu}| = 20$, $p_l^u \in_R [10,100]$, $c_l^{uv} \in_R [10,50]$, $\tau = 20$.

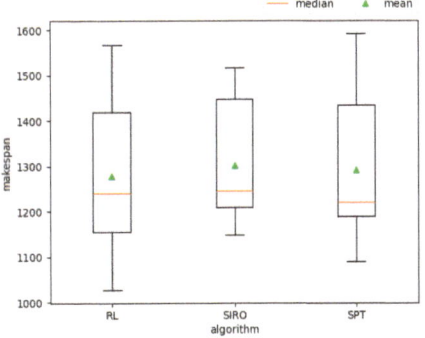

Fig. 2. Evaluation of dispatching rules and an RL scheduling algorithm on the data set by [3]

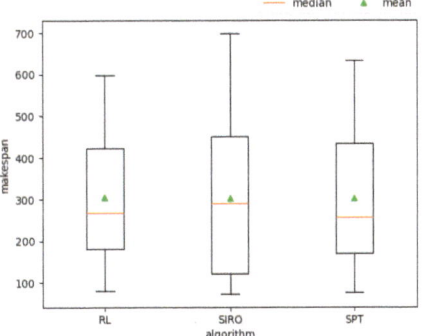

Fig. 3. Evaluation of dispatching rules and an RL scheduling algorithm on the data set by [5]

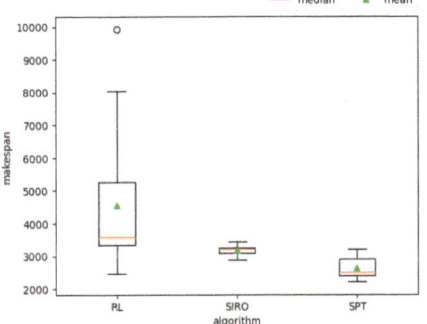

Fig. 4. Evaluation of dispatching rules and an RL scheduling algorithm on the data set by [7]

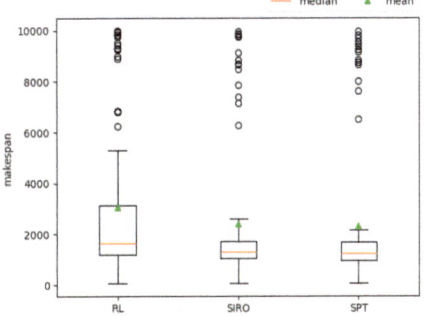

Fig. 5. Evaluation of dispatching rules and an RL scheduling algorithm on the data set by [9]

Figure 6 shows the results evaluated on 100 FJSP instances of the RL scheduling algorithm and the dispatching rules. In contrast to the Figs. 2, 3, 4 and 5 not all boxes for the different scheduling algorithms overlap anymore. The mean performance of the algorithm increases in the order RL, SST+SPT, SPT, SST, SIRO with values 3007.3, 3308.8, 3466.35, 4262.19, 4287.16. While the dispatching rules SIRO and SST have a rather larger box and long whiskers, the RL algorithm as well as the dispatching rules SPT and SST+SPT have a smaller box with shorter whiskers.

5   Discussion

The boxes and whiskers of the different scheduling approaches all overlap in the data sets of [3,5]. Hence, there is no advantage between the scheduling algorithms visible. The results are only slightly better for the data sets of [7,9]. Here one can see an advantage of the dispatching rules SIRO and SPT in contrast to the RL algorithm. These observations suggest, that the RL algorithm is not able to generalize well. One possible reason is that there are too few instances for training of the RL algorithm. Since there is also no advantage between the dispatching rules in Figs. 2 and 3 and only a slight advantage in Figs. 4 and 5, this also suggests that the data sets only have few potential for online scheduling. Since they were created in the 90 s before the fourth industrial revolution started, they were invented for offline scheduling rather than online scheduling. At that time manufacturing systems were rather static and fixed. Thus, offline scheduling techniques where production plans can be computed in advance are generally in favor, since they can achieve more optimal results. With Industry 4.0 dependencies within and among manufacturing systems increased. Hence, flexibility and agility became more important and with this the necessity for online scheduling approaches. This makes these ancient data sets insufficient and outdated for the current research.

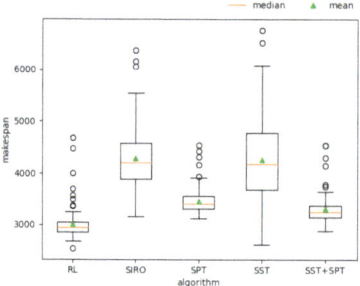

Fig. 6. Evaluation of a RL scheduling algorithm and multiple dispatching rules on the LOS benchmark

For this reason, a new FJSP data set for online scheduling is necessary. By adjusting the values of e.g. the transportation times, setup times or the processing times, we can vary the impact of these factors. This affects the performance of the scheduling algorithm. Hence, we can see a real advantage of the RL algorithm as well as the SST+SPT and SPT dispatching rules over the SIRO and SST dispatching rule in the FJSP as one would expect. This is evident by a visible shift in the boxes and the whiskers, as well as their absolute length. The later is indicating a lower variance in the scheduling results, making the schedules more robust.

The FJSP instance is variably adjustable by its parameter selection. In this way we can scale up the instances itself and examine scheduling algorithms on large scale instances. On the other hand it is also possible to generated numerous amounts of instances for the given parameters. Thus we can generate a large data set with sufficient training instances for e.g. Reinforcement Learning based algorithms to train on. Due to the explicit characterization of the used data set by its parameters the FJSP instances can be recreated. Hence, a comparison of different scheduling algorithms is possible.

6   Conclusion and Future Work

We present a method for the creation of a benchmark data set for FJSP with setup and transportation times. Furthermore, we ensure backwards compatibility to former FJSP data sets. We present results for different dispatching rules on the data set showing its potential for optimization using online scheduling. While the data set was invented for online scheduling, it is not limited to it and can also be used for offline scheduling. Through the potential of the data set for offline scheduling needs to be investigated in further studies. Besides that, future research might focus on the extension of the data set to holonic manufacturing systems and how the holons can be represented in the data set and considered during generation.

Acknowledgement. This work has been funded by the European Commission through the HORIZON-CL4-2022-HUMAN-02 project HumAIne - Hybrid Human-AI Decision Support for Enhanced Human Empowerment in Dynamic Situations (No.101120218).

References

1. Adams, J., Balas, E., Zawack, D.: The shifting bottleneck procedure for job shop scheduling **34**, 391–401 (1988). https://doi.org/10.1287/mnsc.34.3.391
2. Applagate, D., Cook, W.: A computational study of the job-shop scheduling instance. ORSA J. Comput. **3**, 149–156 (1991)
3. Barnes, J., Chambers, J.: Flexible job shop scheduling by tabu search. Graduate Program in Operations and Industrial Engineering, The University of Texas at Austin, Technical Report Series, ORP96-09 (1996)
4. Beasley, J.E.: Or-library: distributing test problems by electronic mail. J. Oper. Res. Soc. **41**(11), 1069–1072 (1990). http://people.brunel.ac.uk/~mastjjb/jeb/info.html
5. Brandimarte, P.: Routing and scheduling in a flexible job shop by tabu search. Ann. Oper. Res. **41**(3), 157–183 (1993)
6. Col, G.D., Teppan, E.: Large-scale benchmarks for the job shop scheduling problem (2021)
7. Dauzère-Pérès, S., Paulli, J.: An integrated approach for modeling and solving the general multiprocessor job-shop scheduling problem using tabu search. Ann. Oper. Res. **70**, 281–306 (1997)
8. Fisher, H.: Probabilistic learning combinations of local job-shop scheduling rules. Ind. Scheduling, 225–251 (1963)

9. Hurink, J., Jurisch, B., Thole, M.: Tabu search for the job-shop scheduling problem with multi-purpose machines. Oper.-Res.-Spektrum **15**(4), 205–215 (1994)
10. Lawrence, S.: Resouce constrained project scheduling: an experimental investigation of heuristic scheduling techniques (supplement). Carnegie-Mellon University, Graduate School of Industrial Administration (1984)
11. Li, Y., Manner, R., Manderick, B.: A genetic algorithm applicable to large-scale job-shop instances. In: Parallel Problem Solving from Nature, pp. 281–290. North-Holland (1992)
12. Mastrolilli, M.: Flexible job shop problem. https://people.idsia.ch/~monaldo/fjsp.html
13. Pfrommer, J., Schleipen, M., Beyerer, J.: PPRS: production skills and their relation to product, process, and resource. In: 2013 IEEE 18th Conference on Emerging Technologies & Factory Automation (ETFA), pp. 1–4. IEEE (2013)
14. Pinedo, M.L.: Scheduling, vol. 29. Springer (2012)
15. Storer, R., Wu, S., Vaccari, R.: New search spaces for sequencing instances with application to job shop 38 (1992) 1495–1509 manage. Science **38**, 1495–1509 (1992)
16. Taillard, E.: Benchmarks for basic scheduling problems. Eur. J. Oper. Res. **64**(2), 278–285 (1993)

Agent-Based Communication for Fault Diagnosis in Skill-Based Production Environments Using Messages Based on I4.0 Language and Asset Administration Shells

Pascal Rübel[1]([⊠]), William Motsch[1], Alexis Bernhard[2], Simon Jungbluth[1], and Martin Ruskowski[2]

[1] Technologie-Initiative SmartFactory KL e.V., Kaiserslautern, Germany
pascal.ruebel@smartfactory.de
[2] German Research Center for Artificial Intelligence (DFKI), Kaiserslautern, Germany

Abstract. Small batch sizes and individualized products are managed with the concept of Cyber-Physical Production Systems (CPPS), enabling flexibility through interchangeable constellations, but increases complexity, especially when dealing with dependencies between decentralized subsystems. To increase the resilience and self-healing capabilities, greater automation of fault detection and diagnosis (FDD) is a key factor. It is a challenge to gather knowledge about faults, as these rarely occur compared to normal behavior. The flexibility in skill-based production systems makes this situation even more difficult. To overcome this challenge, data and knowledge about faults and their context from several Cyber-Physical Production Modules is used, which leads to federated knowledge databases. The knowledge databases are modeled in the Capability-Skill-Service-Fault-Symptom model (CSSFS model). To achieve the goal of high availability, resilience and autonomy of CPPS, automated decision-making for FDD using CSSFS applications is required. Therefore, automatic communication between FDD components is necessary. Therefore, focus of this paper is on the development of a communication scheme, which models participants using Asset Administration Shells and the I4.0 Language to model their interactions to enable automated communication and makes distributed knowledge accessible. To ensure decentralized control of these services, functionalities from several factory levels are encapsulated by Multi-Agent Systems (MAS) that follow a holonic structure.

Keywords: Fault Diagnosis · Holonic Manufacturing Systems · I4.0 Language · Asset Administration Shell

© The Author(s) 2025
K. Alexopoulos et al. (Eds.): ESAIM 2024, LNME, pp. 157–170, 2025.
https://doi.org/10.1007/978-3-031-86489-6_17

1   Introduction

Greater automation of FDD is a key factor in increasing the resilience and self-healing capabilities of manufacturing systems. However, the demand for individualized products requires small batch sizes down to batch size one, that leverages the challenge of FDD. One solution to increase flexibility in factory automation is a modular factory structure and skill-based production. The concept of Cyber-Physical Production Systems (CPPS) enables flexibility through multiple, interchangeable constellations of production subsystems, but on the other hand increases complexity, especially when dealing with the decentralization of models and knowledge bases [1,2]. Even in non-decentralized production environments, it is a challenge to gather experience and knowledge about so-called faults and failures, as these rarely occur compared to the normal behavior of such systems. The increased flexibility in skill-based CPPS makes this situation even more difficult. Individualized tasks and small batch sizes lead to small amounts of individual data and knowledge, which extends the challenge of overall data usage.

One way of overcoming this challenge is to use data and knowledge from several Cyber-Physical Production Modules (CPPM), which lead to federated knowledge databases in which faults are stored with their context. Knowledge databases about faults and their context are modeled in the Capability-Skill-Service-Fault-Symptom model (CSSFS model).

In order to achieve the goal of high availability, resilience and autonomy of CPPS, automated decision-making for FDD using CSSFS applications is required. This requires a communication option that is as autonomous as possible between the components required for FDD.

Against this background, the research objective of this work is on the development of a communication scheme that enables automated communication between FDD components for a decentralized control structure and makes distributed knowledge accessible. Therefore, on the one hand the topic of Asset Administration Shell (AAS) for modelling of CPPMs as communication participants is focused. On the other hand the modelling of the messages between the participants using Industrie 4.0 language is examined. For this purpose, the required FDD components and their tasks and services are defined, which encapsulate distributed forms of knowledge. In order to ensure decentralized control of these services, various functionalities from several factory levels are encapsulated by MAS, that follow a holonic structure. For each task, a heuristic communication scheme is defined to retrieve the corresponding knowledge.

The paper is structured as follows. Section 2 gives an overview of related work in the field. In Sect. 3 an architecture describes the modelling of communication participants and the messages between them using I4.0 components and AAS as well as I4.0 Language. A prototypical implementation of the concept is shown in Sect. 4, whereas Sect. 5 discusses the results and gives an outlook for further research.

2   State of the Art

Following the introduction a brief state of the art in the identified main topics of modular and skill-based production, AAS and I4.0 language as well as fault diagnosis based on multi-agent systems is given in this section.

2.1   Modular, Skill-Based Production

A main element of a modular and flexible production environment is the topic of CPPM, which can provide standardized interfaces for their functionalities and be combined to build CPPS [3]. The CPPMs thus encapsulate their functionality by using the skill-based approach for usage in a flexible production structure [4]. Production skills are part of the information model for capabilities, skills and services (CSS-Model), in which capabilities are described as abstracted functions that are required in a production process and skills as the implementation of such a function and deployed on a specific production resource [5]. Skills can be implemented with industry standards such as OPC UA, to realize a generic interface that can be accessed and executed on higher control levels of a production system [6].

Skills of production resources are useable in a distributed organized control software, enabling their application into an agent-based setup within a MAS [7]. The integration of production skills of production modules into a MAS using resource agents and the design of skills for agent functionality is shown in [8]. In the current contribution, the fault diagnosis in skill-based production environments using agent-based communication is focused.

2.2   Asset Administration Shell and I4.0 Language

The Asset Administration Shell (AAS) is a standard provided by the I4.0 platform and standardized by the Industrial Digital Twin Association with the aim to implement a vendor-independent Digital Twin [9]. In the context of this article, AASs are used as a standardized interface between different subsystems of the CPPS for data exchange, as proposed in [10]. Thereby, AASs extend and represent production assets and leverage them in that way to I4.0 components.

These I4.0 components are accessed by an agent as a part of a holonic multi-agent system (MAS) [11]. This approach uses the agent definition of [12], in which agents are autonomous, problem-solving and goal-driven entities, observing and acting upon an open and dynamic environment. AASs configure and parameterize these agents and represent a standardized interface enabling a heterarchical communication between agents, which enables agents to act and react dynamically to changes in the environment. By the usage of holonic MAS, in the sense that agents can access, dynamically spawn and kill other agents along the ISA-95 factory hierarchy, the AASs of the respective I4.0 components are also hierarchically structured.

Accordingly, the AAS can be used to describe a standardizable structure of interfaces that can also be linked with semantic technologies. For example, the use of knowledge graphs and RDF stores offers potential for applications in the context of Industry 4.0, as presented in our approach in the following sections.

Complete AASs and AAS elements are sent through messages in the I4.0 language. VDI/VDE 2193 standardizes this message format. This guideline is divided into two parts. The first part [13] describes the vocabulary and structure of messages, and the second [14] the semantic interaction protocols. An I4.0 message consists of a message frame and the message content defined by AAS elements [9,10].

First, the frame contains several required elements: the message type, the AAS ID of the sender, the message ID, and a semantic protocol depending on the message type. Specifically, the message type defines the intention of the message, e.g., data inquiries, while the AAS ID and message ID ensure unambiguous tracking and order integrity. In addition, the semantic protocol describes the standardization of message intentions. Secondly, the message content is specified by the interaction elements, for example, the referenced AAS element.[9,10]

The interaction protocol of the I4.0 language is defined through a bidding procedure in which external participants communicate with internal participants (AAS sender/recipient) that communicate in the I4.0 language. This communication is supported by an I4.0 language handler, that transforms data elements into the I4.0 message format.[9,10]

2.3 Fault Diagnosis Based on Multi-agent Systems

In the field of fault diagnosis in manufacturing, the role of MAS is the incorporation of different entities that cover different tasks each. The use of such MAS increases the flexibility in encapsulating multiple resources, tasks and knowledge sources that enable an automated FDD in flexible production systems. For the task of fault diagnosis that is encapsulated in a holon, that is in this work realized by a MAS, a knowledge-based approach is used. The related work in the fields of FDD based on MAS and knowledge based FDD is shown in the following subsection.

For the automation of a fault diagnosis system in power systems, McArthur et al. use a combination of MAS and intelligent systems. Different intelligent decision support systems based on SCADA data and fault records are wrapped within agents to enable automated communication between the systems [15].

In [16], the KARMEN MAS is introduced that covers the task of process monitoring and notification. Component Agents are used that provide operational data of each process component, as well as Condition Monitor Agents that contain logical expressions to evaluate the conditions of the components. The logical expressions are manually defined by the user and can include data of multiple components. After evaluation of the Condition Monitor Agent, a Notification Agent handles the notification and escalation process [16].

The concept of leadership in MAS brings advantages of centralized and decentralized architectures together. Therefore, a set of agents observe sensor values and evaluate them for normal and anomal behaviour. The leader agent is the agent that firstly detects the anomal behaviour. The information of the other agents are aggregated and a spatio-temporal pattern is generated. The patterns are compared with existing fault patterns for fault identification [17, 18].

The safety monitor is realized with a hierarchical MAS including a federated monitoring model and handles multiple system levels. Agents locally reason at subsystem level and collaborate for global reasoning. For monitoring, a model of hierarchical state machines is used for behaviour modelling. This model is combined with a fault propagation model, that consists of numerous fault trees [19–22].

A hybrid reinforcement learning MAS is proposed in [23] focusing the identification of anomalies in industrial microservices. Local outlier determination is executed by agents for each microservice using reinforcement learning. Afterwards the local results are merged to extract global outliers by an intelligent communication strategy.

In the area of knowledge-based fault diagnosis a semantic framework, that describes the manufacturing domain is introduced in [24]. Manufacturing resources, processes and their context are used in combination with system observations to enable stream observations to diagnose situations potentially leading to failures [24]. In [25] connection and component models are developed, that calculate fault symptoms based on components behaviour and their relations. Once a faulty component is identified, the system is reconfigured to bypass it. The new configuration is then fed back as a system update [16]. A general overview of the state of the art of fault diagnosis on CPPS is given by Niggemann and Lohweg in [26].

Although, the stated work in this subsection covers fault diagnosis in CPPS it still lacks the integration of the CSS-model and AAS for knowledge representation and the combination with holonic MAS for automated communication to enable fault diagnosis. Therefore, in this work the focus is on the CSSFS-Model presented in [27] and its usage as a basis for knowledge representation and communication schema introduced in [28] for automated communication. The communication schema is extended with modelling its participants as actice AAS and the message based on I4.0 messages paving the way to higher interoperability, resilience and autonomy of the concept.

3 Structure for Automated Communication for Fault Diagnosis in Flexible Production Systems

For the integration of FDD in flexible production systems, three components are required to be implemented on suitable hierarchy levels of CPPS: A fault detection component, a fault diagnosis component and a knowledge base.

In general, the Fault Detection component triggers a request for a diagnostic task in the event of a deviation from the target behavior. It must therefore transmit the required information to the fault diagnosis component, which queries the knowledge base for possible solutions.

The main task of the fault detection component is the determination of one or multiple faults present in the system. Therefore, a set of symptoms is generated that relates to a fault class. A detailed description of the applied fault detection task can be found in [29]. The fault diagnosis component uses the input of the fault detection component to elaborate analysis on fault classification, root cause analysis and recommendations on the handling of faults. Accordingly, a heuristic communication scheme for the FDD components to execute the different tasks is presented in [28].

In this work, the focus is in the modelling of the communication participants as I4.0 components based on AAS and the messages on I4.0 language (see Sect. 2.2) to enable a knowledge-based approach for fault diagnosis. The properties of a fault and its context are modeled in an information model and made accessible via knowledge graphs. Each fault is modeled at least with its symptoms, the resource on which it occurred, the product produced and the skill that was executed when the fault occurred. This follows the concept of the CSSFS model [27]. Both, the production context and the faults themselves are initially modelled in corresponding AASs to have the knowledge available in a standardized, structured format. To increase searchability and accessibility, these AASs are transformed into knowledge graphs. Doing this, the I4.0 components representation can be used for communication purposes and the knowledge graph representation for analysis and searchability purposes.

In the following, the I4.0 components representation of the communication participants are described. Afterwards, the message structure is described based on the I4.0 language standard.

3.1 Communication Participants

For the usage of I4.0 language, the communication participants need to be modelled as I4.0 components. Therefore, the physical asset is extended with a digital representation to act as a I4.0 component. In this work, type 2 AAS are used as a form of digital representation. AAS of type 2 are reactive AAS, that can be used as an interface. I4.0 components are modelled on the CPPM and CPPS level, whereas multiple CPPM belong to a corresponding CPPS.

On both hierarchy levels submodels are added that contain general information about the asset that are not FDD related in the first place. The mentioned submodels are a minimal configuration and can be extended, depending on the domain and use case. For the FDD use case the modelled information can be used for further analysis.

A digital nameplate submodel, that contains information as usually found on physical nameplates in an interoperable manner, is added to the AAS. Additionally, a submodel for the Bill of Material structure is introduced. This submodel contains the hierarchical structure of industrial equipment that can be composed of different subsystem levels. On the CPPM level, the components of the CPPM are described, whereas on the CPPS level the topology of the corresponding CPPM is modelled. Each CPPS subsystem can have its own AAS. The information about the hierarchical resource structure is required for FDD tasks. Finally, the manufacturing skills that a system can provide is modelled in a submodel. There, the required information about each offered skill is modelled separately in

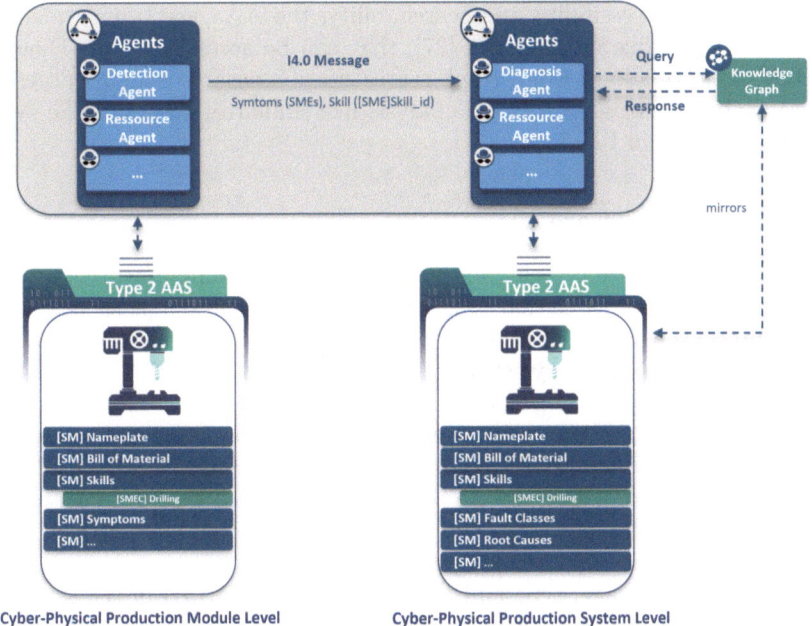

Fig. 1. Overview of the system including participants and messages

a submodel element collection. The information includes an ID for identification purposes and properties for parametrization of the skill.

Additionally to the AAS modelling of the physical components the tasks of FDD need to be integrated in the CPPS hierarchy. The Fault Detection component with its task to monitor executed skills, simulate nominal behavior and generation of symptoms is integrable in a MAS on the CPPM level. A detailed description of the Fault Detection tasks can be found in [29]. Consequently, a submodel is added containing the information about the symptoms that are generated by the symptom generation task. Since the fault detection, and accordingly the symptom generation, is executed on the CPPM level, the symptom submodel extends the CPPM AAS.

Based on the introduced concepts, an integration of a detection agent into a MAS can be proposed, that executes the corresponding tasks and triggers communication with the other agents on CPPM and CPPS level.

Depending on the tasks of the Fault Diagnosis component, further submodels need to be added to extend the information of the AAS. Since information of all CPPMs of the CPPS is available on the CPPS level, the diagnosis is located there. For the tasks of fault classification and root cause determination, a submodel containing fault classes and root causes can both be added. Analog to the CPPM level, a Diagnosis Agent can be suggested for integration in a MAS that handles the tasks and communication.

For a better accessibility and searchability, the AAS are transformed in a knowledge graph, like described in [27], that can be accessed by the Diagnosis Agent. The use of a knowledge graph representation ensures a higher accessibility and enables more sophisticated analysis like similarity analysis. The system overview is shown in Fig. 1.

3.2 Message Structure

After modeling faults and their context, the knowledge graph must be queried to access the stored knowledge that is used to answer the fault diagnosis tasks. Each task requires a separate definition for the queries used. As different scenarios are possible, a hierarchical query schema is developed for each task.

Both, the communication between the fault detection and the fault diagnosis component and the access between the fault diagnosis component and the knowledge graph are modeled according to the I4.0 language.

Using the I4.0 language full AAS or AAS elements can be sent in a standardized format. I4.0 messages consist of a message frame and content, that needs to be defined by a list of AAS elements.

In the frame message, type and ID as well as the AAS ID of the sender are mandatory. Since the message type defines the intention of the message, the message type "fault diagnosis" is used. The AAS ID of the sender is the AAS ID of the CPPM that triggers the interaction with the Diagnosis Agent. The message ID is automatically set and ensures unambiguous tracking and tracing of the messages.

The content of the message is specified by interaction elements in form of AAS or AAS elements. In the FDD use case the symptoms that have been detected as well as information about the resource and the executed skill need to be submitted.

Therefore, the submodel element collections of the generated symptom as well as the submodel element containing the skill ID are added to the message content. The required information about the resource is already part of the message frame in the form of the AAS ID of the sender.

4   Proof of Concept and Results

The system will be implemented as a holonic MAS in the *SmartFactory*KL demonstrator environment. This demonstrates the application of a holonic manufacturing system to encapsulate the intricacies of systems consisting of multiple individual subsystems that are essential for managing complexity in distributed capability-based manufacturing.

Therefore, the entities responsible for FDD along with their respective tasks are defined individually and then integrated into the already existing holonic MAS. Both direct and indirect manufacturing tasks are managed for the CPPS demonstrator, which can ensure a holistic view of the integration of the FDD components in the manufacturing context.

The AAS were modelled using the Eclipse AASX Package Explorer[TM1] and then deployed using Eclipse BaSyx middleware[2]. For the implementation of the holonic MAS in this work SARL language running on Janus Agent and Holonic Platform[3] is used. The communication between the agents follows the standardized structure of Industrie 4.0 language.

4.1 Agent Communication for Fault Detection and Fault Diagnosis

The functionality for fault detection and diagnosis is encapsulated as own agents in the prototypically implemented MAS of the *SmartFactory*[KL] and realized with the Janus Framework, which is an open-source framework for the development of holonic agents, based on the agent-oriented programming language SARL [30]. The agent communication is shown in Fig. 2.

The relevant communication part in the system begins with a request for a production skill execution by a resource agent, which controls the skill execution of a production module. The resource agent informs the fault detection agent about a scheduled skill execution on this production module to prepare a skill monitoring. The fault detection agent monitors the behavior of corresponding production modules during the skill execution and considers the result of this monitoring activity. Detected fault-related data are prepared by the fault detection agent for further analysis by the fault diagnosis agent. The request for fault diagnosis is then used in the fault diagnosis agent to prepare a request to the knowledge base, in which the diagnosis is processed. As a knowledge base, a knowledge graph is used as presented in [27], so that the fault diagnosis agent encapsulates and provides the functionality to build a query for accessing the stored information of this knowledge graph.

4.2 Asset Administration Shell and Agent Based Structure and Communication Implementation

For the implementation, one AAS for the Resource Agent of the running production module and one AAS for the *SmartFactory*[KL] demonstrator are applied. Both in Fig. 3 displayed AASs are modelled and implemented in AAS version 3.

The CPPM agent AAS contains all described standardized submodels such as Nameplate, BOM, Skills, Technical Data, Asset Interface Description, Monitoring, and Symptoms. During the execution of the production module, the detection agent detects new symptoms and adds them to the AAS. As soon as the execution stops, aborts or gracefully ends with a monitoring result, the detection agent sends all newly detected symptoms together with the respective skill ID of the agent to the central diagnosis agent to request a fault diagnosis. The diagnosis agent then checks the detected symptoms via the knowledge

---

[1] https://github.com/admin-shell/aasx-package-explorer.

[2] https://eclipse.dev/basyx/about/.

[3] http://www.sarl.io/runtime/janus/.

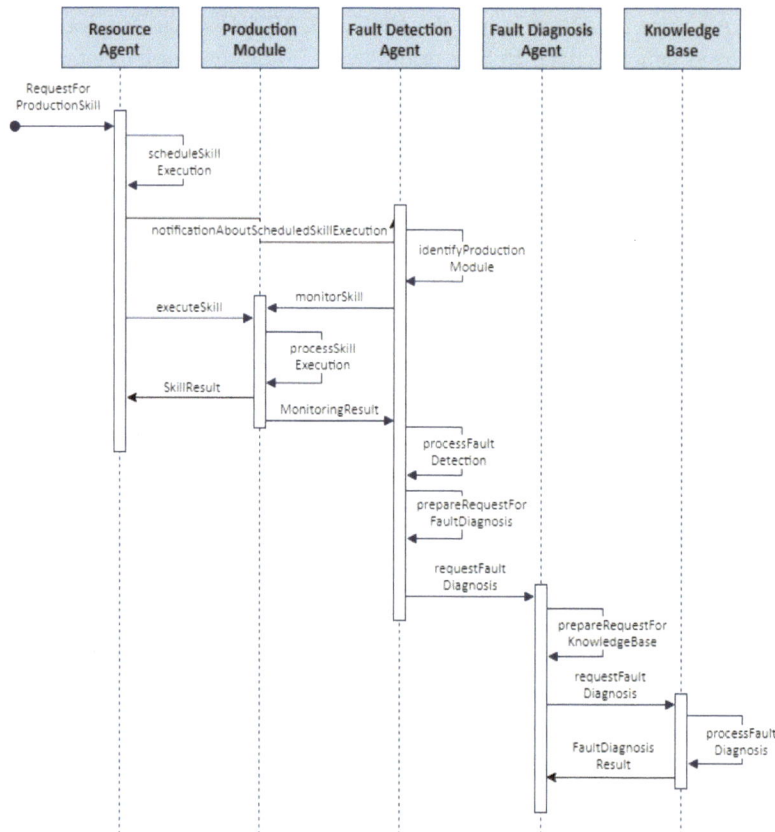

Fig. 2. Sequence Diagram for Agent Communication for Fault Detection and Fault Diagnosis

graph, like describe in [27], and adds references to known faults stored on the CPPS AAS to the symptoms submodel on the production module AAS.

For communication between different agents, messages are sent via interaction protocols. For this purpose, all agents of the MAS have the ability to communicate event-based in the same event space inside the MAS, which is provided by the Janus framework for agent communication. This works perfectly for agents running on the same or closely connected event spaces. This is the case for the CPPM agent and the detection agent, since the CPPM agent dynamically spawns the detection agent on starting a production task and both agents thus share the same agent context. The same situation appears between the CPPS and the diagnosis agent. However, the detection agent and the diagnosis agent might not be deployed on the same device and take part in the same event space. This is why all messages between the agents are embedded into an industry 4.0 message framework, adding more meta-information. Based on the specification

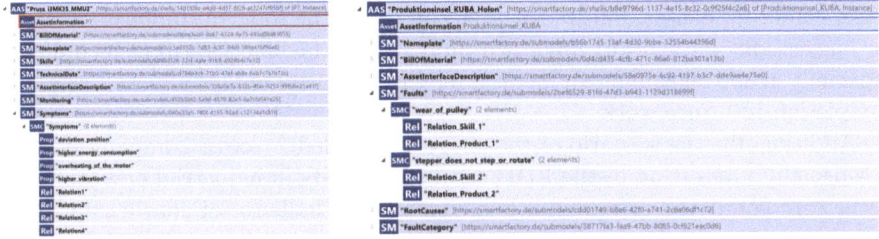

Fig. 3. Asset Administration Shells of a CPPM agent (left side) and the CPPS agent (right side)

of the I4.0 message in [12], a metamodel for I4.0 Messages is defined in Java which contains all mandatory and optional fields of I4.0 messages.

5   Discussion and Conclusion

An agent-based communication structure for the integration of FDD in skill-based production environments is elaborated. The involved participants are described and modelled using type 2 AAS whereas the communication is realized by a MAS and a message structure on the basis of the I4.0 language. Type 2 AAS and the corresponding agents act together similar to the concept of the active type 3 AAS.

In the first step, the FDD tasks were integrated into the different hierarchy levels of CPPS. The Fault Detection Tasks are executed on the CPPM level due to the monitoring and simulation tasks of the behaviour of the CPPM during the execution of manufacturing skills. The tasks are modelled as agent behaviours and encapsulated in a Fault Detection Agent, that covers the communication with other participants. The CPPM as well as the fault symptoms are modelled as AAS and AAS submodels.

The Fault Diagnosis tasks are integrated on the CPPS level, since the access to the information of the whole CPPS is available there. The CPPS is modelled as type 2 AAS as well and required submodels for fault diagnosis are added. A Fault Diagnosis Agent is introduced that manages the internal elaboration of the requested tasks and the communication with participating agents.

The message structure is based on the I4.0 Language standard defined in [9,10]. Therefore, the needed information and its AAS representation is defined and integrated in the message structure.

Finally, the defined structure is prototypically implemented as part of the holonic MAS of the *SmartFactory*[KL] demonstrator environments. Besides the proof of concept the basic interactions between the existing agents in the MAS are shown. In this work, the focus was on the detection and diagnosis of a priori known faults. In this case, it is possible to get existing matches of instances, that are already in the knowledge base. In the case of a priori unknown faults, further analysis needs to be executed to elaborate suggestions on possible fault classes and root causes. Future research can tackle the topic of similarity analysis for suggestions on a priori unknown fault cases. Machine Learning techniques, especially graph learning, can be used for the similarity analysis as well as graph completion in future work.

Additionally, the scenario of the usage of knowledge across company borders is especially interesting in the FDD case, since faults in industrial environments occur rare compared to nominal behaviour. The challenge of sharing data and knowledge of multiple companies using data spaces can be developed in future research.

Acknowledgement. This work was supported by the German Ministry for Economic Affairs and Climate Action by the project "Aufbau eines Datenraums für die ausrüstende Industrie – die Ausrüster von Fabriken weltweit" (Factory-X) under grant 13MX001ZU.

References

1. Hermann, J., Rübel, P., Birtel M., Mohr, F., Wagner, A., Ruskowski, M.: Self-description of cyber-physical production modules for a product driven manufacturing system. Procedia Manuf. **38**, 291–298 (2019)
2. Ruskowski, M., et al: Production Bots für Production Level 4: Skillbasierte Systeme für die Produktion der Zukunft. ATP Magazin **62**(9), 62–71 (2020)
3. Kolberg, D., et al.: SmartFactoryKL System Architecture for Industrie 4.0 Production Plants; Whitepaper SF-1.2, 4; SmartFactoryKL: Kaiserslautern, Germany (2018)
4. Bergweiler, S., Hamm, S., Hermann, J., Plociennik, C., Ruskowski, M., Wagner, A.: Production Level 4-Der Weg zur Zukunftssicheren und Verlässlichen Produktion; Whitepaper SF-5.1; SmartFactoryKL: Kaiserslautern, Germany (2022)
5. Diedrich, C., et al.: Information Model for Capabilities. München, Germany, Skills & Services; Fraunhofer-Gesellschaft (2022)
6. Dorofeev, K., Zoitl, A.: Skill-based engineering approach using OPC UA programs. In: 2018 IEEE 16th International Conference on Industrial Informatics (INDIN), pp. 1098–1103. IEEE (2018)
7. Ruskowski, M., et al.: Production Bots für Production Level 4: Skill-basierte Systeme für die Produktion der Zukunft. ATP Mag. **62**, 62–71 (2020)
8. Sidorenko, A., Motsch, W., Van Bekkum, M., Nikolakis, N., Alexopoulos, K., Wagner, A.: The MAS4AI framework for human-centered agile and smart manufacturing. Front. Artif. Intell. **6**, 2023 (2023)
9. Federal Ministry for Economic Affairs, and Energy (BMWi), Industrial Digital Twin Association: Details of the Asset Administration Shell Part 1 - The exchange of information between partners in the value chain of Industrie 4.0 (Version 2.0.1). https://www.plattform-i40.de/IP/Redaktion/DE/Downloads/Publikation/Details_of_the_Asset_Administration_Shell_Part1_V2.pdf?__blob=publicationFile&v=6 (2019)
10. Federal Ministry for Economic Affairs, and Energy (BMWi), Industrial Digital Twin Association: Details of the Asset Administration Shell Part 2 - Interoperability at Runtime - Exchanging Information via Application Programming

Interfaces (Version 1.0RC02). https://www.plattform-i40.de/IP/Redaktion/EN/Downloads/Publikation/Details_of_the_Asset_Administration_Shell_Part2_V1.pdf?__blob=publicationFile&v=1 (2021)

11. Jungbluth, S., et al.: Dynamic replanning using multi-agent systems and asset administration shells. In: 2022 IEEE 27th International Conference on Emerging Technologies and Factory Automation (ETFA), pp. 1–8. IEEE (2022)

12. Leitão, P., Karnouskos, S.: Industrial agents: emerging applications of software agents in industry. Elsevier (2015)

13. VDI/VDE 2193 Blatt 1 - Sprache für I4.0-Komponenten - Struktur von Nachrichten (2020)

14. VDI/VDE 2193 Blatt 2 - Sprache für I4.0-Komponenten - Interaktionsprotokoll für Ausschreibungsverfahren (2020)

15. McArthur, S.D.J., Davidson, E.M., Hossack, J.A., McDonald, J.R.: Automating power system fault diagnosis through multi-agent system technology. In: Proceedings of the 37th Annual Hawaii International Conference on System Sciences, p. 8. IEEE, Big Island, HI, USA (2004)

16. Bunch, L., et al.: Software agents for process monitoring and notification. In: Proceedings of the 2004 ACM Symposium on Applied Computing, pp. 94–100. ACM, Nicosia Cyprus (2004)

17. Mendoza, B., Xu, P., Song, L.: A multi-agent model for fault diagnosis in petrochemical plants. In: 2011 IEEE Sensors Applications Symposium, pp. 203–208. IEEE, San Antonio, TX, USA (2011)

18. Mendoza, B., Xu, P., Song, L.: A multi-agent model with dynamic leadership for fault diagnosis in chemical plants. In: Abraham, A., Corchado, J.M., González, S.R., und De Paz Santana, J.F. (hrsg.) International Symposium on Distributed Computing and Artificial Intelligence, pp. 19–26. Springer Berlin Heidelberg, Berlin, Heidelberg (2011)

19. Dheedan, A.A.: Multi-agent on-line monitor for the safety of critical systems (2012). https://doi.org/10.5281/ZENODO.1075613

20. Dheedan, A.A.: Distributed on-line safety monitor based on safety assessment model and multi-agent system (2012)

21. Dheedan, A., Papadopoulos, Y.: Multi-agent safety monitor. In: IFAC Proceedings, vol. 43, pp. 84–89 (2010). https://doi.org/10.3182/20100701-2-PT-4011.00016

22. Dheedan, A., Papadopoulos, Y.: Model-based distributed on-line safety monitoring. In: The Third International Conference on Emerging Network Intelligence (EMERGING 2011), Lisbon, Portugal, pp. 1–7 (2011)

23. Belhadi, A., Djenouri, Y., Srivastava, G., Lin, J.C.-W.: Reinforcement learning multi-agent system for faults diagnosis of mircoservices in industrial settings. Comput. Commun. **177**, 213–219 (2021). https://doi.org/10.1016/j.comcom.2021.07.010

24. Giustozzi, F., Saunier, J., Zanni-Merk, C.: A semantic framework for condition monitoring in Industry 4.0 based on evolving knowledge bases. **15**, 583–611 (2024). https://doi.org/10.3233/SW-233481

25. Diedrich, A., Balzereit, K., Niggemann, O.: First approaches to automatically diagnose and reconfigure hybrid cyber-physical systems. In: Beyerer, J., Maier, A., und Niggemann, O. (hrsg.) Machine Learning for Cyber Physical Systems, pp. 113–122. Springer Berlin Heidelberg, Berlin, Heidelberg (2021)

26. Niggemann, O., Lohweg, V.: On the diagnosis of cyber-physical production systems. State-of-the-Art Res. Agenda (2015). https://doi.org/10.24406/PUBLICA-FHG-388129

27. Rübel, P., Moarefvand, N., Motsch, W., Wagner, A., Ruskowski, M.: Enabling fault diagnosis in skill-based production environments. In: 2023 IEEE 28th International Conference on Emerging Technologies and Factory Automation (ETFA), pp. 1–8. IEEE, Sinaia, Romania (2023)
28. Rübel, P., Jungbluth, S., Motsch, W., Ruskowski, M.: Automated communication for fault diagnosis in flexible production environments. In: Proceedings of the Flexible Automation and Intelligent Manufacturing International Conference, Accepted. Taichung, Taiwan (2024)
29. Rübel, P., Motsch, W., Schäfer, H., Ruskowski, M.: On bringing fault detection to skill-based production. In: Proceedings of the 1st European Symposium on Artificial Intelligence in Manufacturing (ESAIM), accepted. Kaiserslautern, Germany (2023)
30. Rodriguez, S., Gaud, N., Galland, S.: SARL: a general-purpose agent-oriented programming language. In: Proceedings of the 2014 IEEE/WIC/ACM International Joint Conferences on Web Intelligence (WI) and Intelligent Agent Technologies (IAT), Warsaw, Poland, 11-14 August 2014; IEEE: Piscataway, NJ, USA, vol. 3, pp. 103–110 (2014)

Decision Support System (DSS)
for Manufacturing Engineering of Cans Rolling

Ander Martín[1]([✉]), Mariluz Penalva[1], Fernando Veiga[2], Cristina Ruiz[3],
and Víctor Martínez[3]

[1] Affiliation a TECNALIA, Basque Research and Technology Alliance (BRTA),
Parque Científico y Tecnológico de Guipúzcoa, 20009 Donostia-San Sebastián, Spain
`{ander.martin,mariluz.penalva}@tecnalia.com`
[2] Departamento de Ingeniería, Universidad Pública de Navarra, Edificio Departamental Los
Pinos, Campus Arrosadía, 31006 Pamplona, Navarra, Spain
`fernando.veiga@unavarra.es`
[3] IDESA Ingeniería y Diseño Europeo, PCTG. Edificio Félix Herreros, 33203 Gijón, Spain
`{cristina.ruiz,victor.martinez}@idesa.net`

Abstract. Decision Support Systems (DSS) can help factory workers in the
decision-making step of multiple tasks. In digital factories, these systems make
use of data towards a human-centered manufacturing. Rolling of large and thick
plates into cans is a common practice in the metal forming industry to fabricate
pipes or tanks. The process is adjusted by trial and error with a high level of operator intervention. Furthermore, only a small number of cans are identical. The
objective of this work is to prescribe, by means of a DSS, the process parameters
to be applied by the operator in the machine to optimize the can fabrication. The
development of the DSS involved several steps, including firstly signal preprocessing and classification and then data extraction, aggregation, and regression in
a multi-stage prediction framework. A significant use of domain knowledge for
a data-centric solution contributes to the quality of the recommendations and the
ability to organize and transfer know-how among operators.

Keywords: Decision Support System · machine learning · data-centric
regression · classification · metal forming · data aggregation · domain
knowledge-based feature extraction

1 Introduction

Decision Support Systems (DSS) are expected to play a relevant role in the implementation of the Industry 5.0 paradigm, which promotes to maximize the benefits of factory
digitalization under a human-centered approach. DSSs have indeed the potential to provide workers with the necessary information in the decision-making step in operations
design, execution and evaluation. Up to date, DSSs have been proposed for multiple
manufacturing tasks, from the design and optimization of products and processes to the
diagnosis of either processes or machinery [1–3]. With the rise of the digital factory,

© The Author(s) 2025
K. Alexopoulos et al. (Eds.): ESAIM 2024, LNME, pp. 171–179, 2025.
https://doi.org/10.1007/978-3-031-86489-6_18

new options open up for DSSs through the increasing amount of data which are more and more readily available [4]. Decision-making models can benefit not only from an increased amount of data but also from the integration of artificial intelligence techniques that can enhance the results of the decision-making process.

Rolling of large and thick plates into cans is an apparently simple manufacturing process but yet challenging. Each can is almost unique, with continuous variations in the material and dimensions. The process requires from important manual intervention and an error in the setting of the process parameters can be fatal due to the high value of each part. Numerical models can support the phase of parameters setting [5] but, in practice, they are economically unaffordable for a close to one-of-a-kind production. Furthermore, numerical models do not take into account the machinery evolution under heavy working conditions. Being manually operated machines, the operator becomes an additional factor with criteria varying between different operators. Difficulties in training skilled operators is another challenge which cannot be skipped either.

In this paper, a DSS for cans rolling based in machine data, domain knowledge, the application of machine learning techniques and an inference learning strategy is presented. The solution has been designed for the ease of the setting of the rolling parameters as well as to be a training tool for new machine operators. The development of the DSS involved several steps that are explained in the following sections: Methodology including domain contextualization, materials and data, domain-knowledge based feature extraction, data aggregation and predictive regressions; validation, and conclusions.

2   Methodology

2.1   Domain Contextualization

The analyzed can rolling process is made in a three-roller bending machine. This essentially consists of a metallic plate, an upper front roller, and two lower rollers (front and rear). The position of the front roller is fixed, as shown in Fig. 1 and the distance between the upper roller and the lower rollers is adjusted according to the thickness of the plate. These two rollers apply no bending pressure on the plate; they just support it and rotate allowing it to pass through without slipping.

The operation starts by inserting one end of the plate into the bending machine. Then the plate is bent into a curve by adjusting the position of the rear roller, while the front rollers rotate making the plate advance. Curvature is given in steps of about 1 m, and for each step a multi-pass strategy is carried out. Each pass requires at least one forward movement to overcome the yield point and one backward movement to achieve the curvature. This process is repeated until the curved plate reaches tolerances, checked by the operator with a template. The multi-pass forming bases on "trial and error" and is very time consuming every time a new can design must be fabricated. Furthermore, the operator is expected to set the right parameters with no other support than his own know-how and despite the continuous changes of material and design.

The process parameters to define are: 1) position of the wedges (left and right) that adjust the gap between the front rollers, 2) position of the rear roller for the forward movement related to the overcome of the yield point, 3) range (min, max) of positions of the rear roller for the curving.

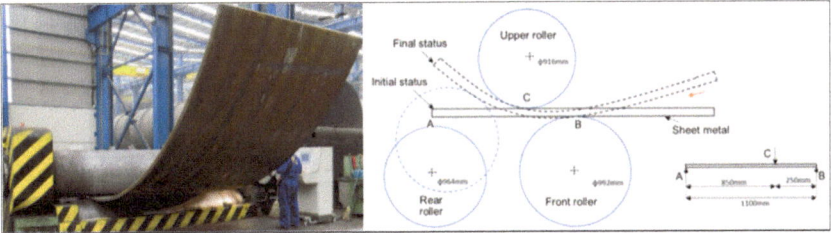

Fig. 1. Three-roll bending picture and diagram.

2.2 Materials and Data

The curving machine is equipped with a PLC (Programmable Logic Controller) signal acquisition system, though the recording must be activated manually by the operator. The data recorded are as follows: 216 files of process signals (time-series) out of which 113 were correctly recorded and 103 were rejected due to errors during the manual activation of the recording. Out of the 113 correct files, there are 34 different can design combinations of diameter, thickness and material. Iterations in the design range from just one sample up to 35. Each can recording provides 98 variables (both analogue and binary) sampled at 2 Hz. The present work is based on data from steel cans with diameters ranging between 1840 and 7730 mm and thicknesses between 9,5 and 79,4 mm.

In addition to the PLC data, there is a knowledge base based on the operators' experience that provides parameters recommendation for a reduced number of previously manufactured cases.

Based on the results of a previous related work [6] the signals of interest for parameters setting were already identified. Concerning the analogue signals, those of interest are: the rotation speed of the top (main) cylinder, the position and pressure of the hydraulic cylinder moving the rear roller, the timestamp and the variables that fix the gap between the top and bottom front rollers (position of the left and right wedges). In terms of binary signals, that of the main motor actuator provides useful information to differentiate the active process (rolling) from the non-active one (no rolling).

2.3 Domain-Knowledge Based Feature Extraction

It has been demonstrated [7, 8] that in the manufacturing domain data-centric models, which take into account domain knowledge for data pre-processing and feature extraction, show better prediction outcomes and help to better identify the most influencing physical variables.

In the rolling process under study, manual operation of the machine is source of highly uneven signals due to high variability in the process execution (e.g. plate curving or manipulation, process in standby) or errors when activating the recording. In order to mitigate the impact of these irregularities on the feature extraction, the reception of the signals was automated following domain knowledge criteria.

First of all, a human in the loop strategy decides whether each signal is appropriate for feature extraction and which portion of it is meaningful for it. To do so, the three different stages of the process are checked on the signal: (1) forming of the 1st head, (2) forming of the 2nd head plus 1/4 of the plate, and (3) forming of the middle part of the plate from the formed 1/4 up to the 1st head (Fig. 2).

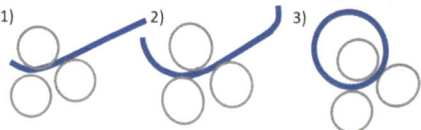

Fig. 2. Three main stages of the rolling process: (1) forming of the 1st head, (2) forming of the 2nd head plus ¼ of the plate, (3) forming of the remaining plate from the ¼ up to the 1st head.

Between stages (1) and (2), there is a timeframe when the rear roller applies no pressure on the plate, it turns down and the gap between the front rollers increases to take out the plate and reposition it for the upcoming stage. This can be tracked through the rear roller position, which drops below an identified threshold (350 mm). Signals with more than two drops below threshold are rejected since that means they contain the rolling of more than one can.

On the other hand, curving is made at regular steps which are preceded by a forward movement of the plate of 1 to 1.5 m length. This sequence keeps tracked through the upper main roller rotation signal with the forward movement showing regular duration (about 30 s). Along the transition between stages (2) and (3), the plate also moves forward, though from the curved section (2nd head plus 1/4) in stage (2) all the way to the front 1st head to start with stage (3). The movement takes more than twice compared to that associated to curving, as shown by the green arrow in Fig. 3. This allows to split stage (2) from stage (3).

After dividing and manually labelling the signals into two classes:1) suitable for process parameter identification (with several peaks) and 2) non-suitable (without peaks); a gradient boosting classification model was trained to recognise the shape of the signals from the following features: the number of identified peaks in the rear roller position signal within the operating range, the mean value of these peaks and the area under the signal divided by the length of the signal, all normalized. Signals suitable for process parameter identification are those with segments with a sequence of peaks as saw teeth, such as in the first two stages in Fig. 3. The target feature for the displacement of the rear roller to the bending position is related to the maximum value of the peaks.

A total of 318 signal segments were used, divided as per the previous description. The model was created to identify the appropriate segment of the signal for parameter recognition.

To evaluate the performance of the model, 5-fold cross-validation was used. In this process, the data is divided into 5 groups or "folds". The model is then trained and tested 5 times, each time using a different fold as the test set (20%) and the remaining folds (80%) as the training set. This approach provides a more robust measure of the model's performance by averaging the results over multiple runs (Table 1).

Data for creating this model were slightly unbalanced (40% vs 60%). Data balancing methods were tried, and results did not show relevant differences, except in detecting

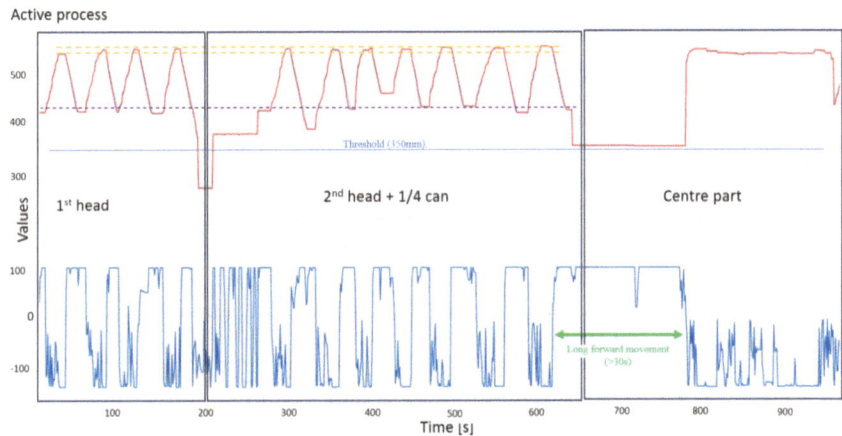

Fig. 3. Rolling process signals (red: position of the rear roller, blue: rotation of the main cylinder) divided into stages and features extracted to represent the position of the rear roller (orange and purple).

Table 1. Results of the classification model.

Metric	Fold 1	Fold 2	Fold 3	Fold 4	Fold 5	Average
Accuracy	0.8714	0.9143	0.9143	0.9565	0.8551	0.9023
Precision	0.8787	0.9275	0.9275	0.9569	0.8559	0.9093
Recall	0.8714	0.9143	0.9143	0.9565	0.8551	0.9023
F1-Score	0.8705	0.9141	0.9141	0.9565	0.8548	0.9020

false positive which is critical in this application. For this reason, an unbalanced dataset was finally chosen.

Once the segments of the signals in which rolling parameters can be robustly extracted was identified, feature extraction was carried out. The displacement of the rear roller for forward movement and for bending were obtained by identifying peaks and valleys in a range of the active process signal where curving occurs (orange and purple lines in Fig. 3). For the position of the wedges that adjust the gap between the rollers, the mode value of the corresponding signal segment during the non-active process is used.

Thanks to this procedure, signals containing valuable information are identified and their features extracted automatically. Those features will be used as the labels for the predictive regressors of the DSS.

2.4 Data Aggregation

Another crucial step in the work was to relate the process parameters obtained from signals with dimensions and material properties. The material design information is stored at the Enterprise Resources Planning (ERP) platform. Each material, each plate, has an ID number and the signal fields are stored with the ID of the corresponding plate. Thanks to these codes, it is possible to link the information from the signal with the characteristics of the plates.

The correlation between material properties and signal characteristics was analysed. Despite the limited number of designs produced (only 34 different combinations of material, thickness and radius) the results pointed out that large curvature radii are associated to small displacements of the rear roller. Regarding the width of the plate, no trend was observed in relation to the position of the rear roller, only to say that the width is related to the weight and heavier plates result in higher pressures. In terms of material, it shows an influence in the rolling process, but the exact effect in the parameters is not clear with the data available at this point.

The descriptors obtained from the signals by feature extraction and the information about the design of the cans were merged into spreadsheet files.

2.5 Predictive Regressions

Once the data were organised, regression models were proposed: on the one hand, to follow the logic of the workers, the nearest neighbour's strategy was considered, but the data sample did not cover the area of use in a homogeneous way, and it could happen that the nearest neighbours were far away or very different. Other types of regression models were also considered, such as decision trees, but they were discarded because of the limited data. There are only 34 different combinations of design.

Finally, with the data available, multi linear regression models were chosen. Table 2 shows selected inputs and outputs for each parameter recommendation. As the elastic limit, ultimate strength and elongation are related to the material, it is sufficient to enter the thickness, radius and material to obtain a recommendation from decision support system in rolling design. The recommendation for the rear roller displacement for forward movement has not been considered because this value is not relevant for small diameter cans, the risk of overbending is much lower and the value is not repeated. Leaving aside small diameter cans, there are not enough samples to consider a regression model for rear roller displacement for forward movement.

5-fold cross validation was used to evaluate the performance of the selected algorithms. Due to the limited data, the number of samples in each fold is limited, and each validation is highly dependent on the data, e.g. for the position of the wedges the fold 4 is not representative for the data set, or for the displacement of the rear roller the fold 1 is less representative for the data set. Table 3 shows the results of the cross validation for the regressions.

To put the results in context, it should be noted that for the same references, the parameters show differences from the median due to manual operation. This is shown in Table 4, where the parameters are numbered as in the previous table. The results are positive considering the lack of repetitively of the process but need to be validated with more data.

Table 2. Inputs used for the regression models of the rolling parameters.

Parameter (Outputs)	Inputs
1) Position of left wedge to adjust the gap between rollers	Thickness
2) Position of right wedge to adjust the gap between rollers	Thickness
3) Displacement of the rear roller for curving (max)	Thickness, Radius, Material properties and estimation from a parabolic regression of the operators' data
4) Displacement of the rear roller for curving (min)	Thickness, radius, material properties and regression result for displacement of the rear roller for curving (max)

Table 3. Results of the regression models of the rolling parameters.

Metric	Fold 1	Fold 2	Fold 3	Fold 4	Fold 5	Average
1) Position of left wedge to adjust the gap between rollers						
Mean Squared Error (MSE)	3.78	2.19	3.01	24.06	5.26	7.66
Mean Absolute Error (MAE)	1.63	1.29	1.45	2.93	1.94	1.85
R-squared (R^2)	0.86	0.94	0.91	−0.67	0.69	0.55
2) Position of right wedge to adjust the gap between rollers						
MSE	6.04	2.60	3.87	9.28	7.30	5.82
MAE	1.78	1.34	1.63	2.38	2.21	1.87
R^2	0.83	0.93	0.87	−0.26	0.65	0.60
3) Displacement of the rear roller for curving (max)						
MSE	229.86	98.20	15.54	56.39	192.06	118.41
MAE	12.84	7.64	2.96	6.13	12.04	8.32
R^2	0.39	0.95	0.99	0.85	0.89	0.81
4) Displacement of the rear roller for curving (min)						
MSE	177.31	245.87	100.72	233.74	180.55	187.64
MAE	11.05	8.52	8.97	11.20	11.47	10.24
R^2	0.50	0.88	0.91	0.70	0.83	0.77

Table 4. Averaged differences to each design median.

Parameter	1	2	3	4
Averaged differences to each design median (mm)	1.24	1.35	4.82	7.59

3   Limitations

These models need to be tested in a real production environment and with more data for the following reasons: a) The feature extraction has been done using real process signals and as the process follows a multi-pass trial-and-error strategy, the results for the same reference are not strictly repeatable; b) also, there is more than one worker using the rolling machine and each of the machine operators have their own skills, some are more cautious and others, due to their experience, take more risks in the trial-and-error strategy; c) the new responsibility for the recording of the process signals lies on the operators and they sometimes forget to start or stop the recording at the right time, which affects the results; d) sometimes the head of the plates is formed in advance and this step of the process is excluded from the recorded signals and other times it is included; e) the amount of available data to build up the models up to date is low f) and last but not least, parameters extraction is based on a AI model and it could produce error propagation. These aspects make the feature extraction less reliable and the results less repeatable, so it is crucial to record the signal in the right way and validate the models in real environment. More data is needed to obtain a robust solution and an industrial validation, but the solution is ready for retrains and read knew data.

4   Conclusions

This work presents a Decision Support System (DSS) for the manufacturing engineering of cans rolling, aiming to optimize the process reducing the trial & error current practises. Despite the limited data and the inhomogeneity of the signals, good results have been obtained in prescribing the parameters, although validation is pending. The DSS is designed to learn incrementally as it is used, updating the data from machine signals and recommendation models accordingly.

A web application has been developed to roll out of the development. It will allow to update the data and the recommendation models. Workers will be able to search the parameters from the workshop before facing a new can.

While the tool is prepared to retrain the models, introducing a criterion for this update is necessary. Techniques for evaluating concept drift could be beneficial in this context, ensuring the models remain accurate and relevant over time. This could be a valuable addition to the system, but due to the limited data it is not addressed in this work.

Overall, this work represents a significant step forward in the application of artificial intelligence and data-centric regression techniques in low volume and highly manual manufacturing operations.

Funding. The authors acknowledge the funding from the European Union's Horizon 2020 research and innovation programme under grant agreement No. 958303.

References

1. Mumali, F.: Artificial neural network-based decision support systems in manufacturing processes: a systematic literature review. Comput. Ind. Eng. **165**, 107964 (2022). https://doi.org/10.1016/j.cie.2022.107964
2. Stavropoulos, P., Papacharalampopoulos, A., Stavridis, J., Sampatakakis, K.: A three-stage quality diagnosis platform for laser-based manufacturing processes. Int. J. Adv. Manuf. Technol. **110**, 2991–3003 (2020). https://doi.org/10.1007/s00170-020-05981-9
3. Hopmann, C., Sasse, J.: Development of a decision support system for profile extrusion. AIP Conf. Proc. **3012**, 020015 (2024). https://doi.org/10.1063/5.0192051
4. Benjamin James, R., Stockinger, M.: Digitalization and digital transformation in metal forming: key technologies, challenges and current developments of industry 4.0 applications. In: XXXIX. Verformungskundliches Kolloquium, Montanuniversität Leoben, Lehrstuhl für Umformtechnik, pp. 13–23 (2020)
5. Bhujangrao, T., Veiga, F., Penalva, M., Costas, A., Ruiz, C.: Three-dimensional finite element modelling of sheet metal forming for the manufacture of pipe components: symmetry considerations. Symmetry **14**, 228 (2022). https://doi.org/10.3390/sym14020228
6. Penalva, M., et al.: Application-oriented data analytics in large-scale metal sheet bending. Appl. Sci. **13**, 13187 (2023). https://doi.org/10.3390/app132413187
7. Mende, H., Frye, M., Vogel, P.-A., Kiroriwal, S., Schmitt, R.H., Bergs, T.: On the importance of domain expertise in feature engineering for predictive product quality in production (2023). https://doi.org/10.24406/publica-2405
8. Sabatakakis, K., Bourlesas, N., Bikas, H., Papacharalampopoulos, A., Stavropoulos, P.: Laser Welding of dissimilar cell tabs: extracting physics semantics from infrared (IR) emissions as process monitoring data. Procedia CIRP **121**, 222–227 (2024). https://doi.org/10.1016/j.procir.2023.09.251

AI Based Solutions for Manufacturing Mass Customization

Luis Usatorre$^{(\boxtimes)}$ ⬥, Paula Morella ⬥, Iñigo Sedano ⬥, Sergio Clavijo ⬥, and Asier Aguayo ⬥

Fundacion Tecnalia Research and Innovation, Member of the Basque Research Technology Alliance (BRTA), Mendaro, Spain
`luis.usatorre@tecnalia.com`

Abstract. This paper analyses how to solve the challenges in the implementation of Mass Customization in manufacturing using Artificial Intelligence agents/services/tools. Considering that humans alone cannot cope with mass customization due to the huge amount of information, it is required AI based solutions that help humans to take decisions. We consider that those AI based solutions must communicate with other AI based solutions in order to obtain a holistic improvement (this is the Multi Agent System concept). More in detail, this paper addresses how to solve the challenges identified when AI based agents use external data coming from outside the company, so a Data Space to guaranteeing a secure data transaction and data ownership and sovereignty is required.

This paper presents the solutions implemented in several projects to address the challenges created by the requirements on a-the implementation and scalability of AI based solutions in Manufacturing, b-for the implementation of multi-Agent AI-based systems and for c-implementing Data Spaces.

Keywords: Artificial Intelligence · Manufacturing · Data Spaces · Multi Agent System (MAS) · Value Chain (VC)

1 Introduction

The future of manufacturing in Europe is poised towards mass customization, driven by market variations, supply disruptions, and changes in auxiliary resources (such as energy and water) due to demographic shifts and climate change. While this scenario promises faster fabrication lead times, reduced time-to-market efforts, and improved customer customer satisfaction [1], it also presents significant challenges related to flexibility, responsiveness, and sustainability.

To address these challenges, this paper presents a solution enhancing inter-organizational information exchange within a Multi-Agent System (MAS) framework. The approach involves integrating various AI-based services for manufacturing and auxiliary processes. Additionally, standardized data, models (using the Asset Administration Shell, AAS) is employed for efficient data storage, while new communication protocols and connectors (such as the International Data Spaces, IDS) facilitate seamless data transactions.

K. Alexopoulos et al. (Eds.): ESAIM 2024, LNME, pp. 180–193, 2025.
https://doi.org/10.1007/978-3-031-86489-6_19

AI-based services, acting as agents, play a crucial role in optimizing manufacturing process parameters, maintenance, and scheduling. These agents consider multi-objective requirements and adapt to the use of novel recycled materials, aligning with circularity objectives. The result is improved accuracy and responsiveness in manufacturing operations.

But when dealing with supply or value chain or when addressing the whole product lifecycle, the data that AI based agents use is coming from many different sources outside the company. In order to guarantee a secure data transaction, guaranteeing the data ownership and sovereignty it is required to implement a Data Space (DS) that permits the seamless data transaction amongst numerous stakeholders.

With this in mind, this paper analyses the requirements that currently exist for these issues, how they interrelate and how they can be addressed. This analysis is guided by the research questions (RQs) presented below:

- RQ1. – What are the requirements for the implementation and scalability of AI based solutions in Manufacturing and in Circularity?
- RQ2. – What are the requirements for a Multi-Agent AI-based system?
- RQ3. – What are the requirements implementing Data Spaces?

As a result, this paper present the challenges addressed when following the requirements and aims also to answer the question *What do Data Spaces (DS) bring to AI applications?*. This question will be discussed as one of the paper conclusions.

The paper is organized as follows:

- Section 2 introduces the theoretical background about DS and AI.
- Section 3 describes the requirements found in previous projects and the experience acquired overcoming them.
- Finally, the paper ends with conclusions and proposals for future research in Sect. 4.

2 Theoretical Background

2.1 Background on AI in Manufacturing and Circularity

In the EU, manufacturing systems are advanced yet not fully optimized for flexibility and resiliency, particularly in terms of energy and water use. Traditional approaches focus mainly on efficiency and cost, often overlooking the significance of optimizing machine parameter configurations for resource consumption.

There is a gap in integrating these considerations into production planning, which leads to missed opportunities for reducing energy and water usage. Additionally, most existing systems are not designed to dynamically adapt machine settings based on real-time operational data, which could significantly enhance sustainability and operational efficiency. In projects like DIGIPRIME and MASA4AI, the problems of planning and reconfiguration of machines have been addressed to make them more efficient, but this has been done by combining different agents within a factory and without considering data or results coming from outside or essential resources such as water or energy.

Current manufacturing systems in the EU are optimized mainly for efficiency within single entities, lacking in comprehensive Value Chain (VC) integration [2]. This narrow focus results in operational scoreboards and VC assessments do not provide a complete picture of the supply chain's dynamics and risks. Additionally, existing solutions fall short in considering the stochastic behaviour and risks of the organization's VC, leading to gaps in seamless system performance and optimal resource utilization.

Several studies have proposed different indicators to assess VC, but there is no standardized set of indicators to measure it. Regarding the development of digital tools to assess the VC, the majority of identified tools focused on the human domain [3], while others, such as DHL [4], have worked on a transportation process tool and only for external factors (natural disasters, meteorology, accidents...).

In order to provide a concise overview of the artificial intelligence technologies that will be used in our services, these technologies will be grouped into 4 areas:

- AI-1: Data Analysis. This group includes techniques that enable data analysis to enhance users' understanding of its functioning. It encompasses techniques such as data cleaning, univariate analysis, bivariate analysis, statistical inference tests, outlier identification and active learning.
- AI-2: Machine and Deep Learning. This group encompasses various algorithms used in classification and regression problems, such as neural networks, convolutional networks, random forests, Support vector machine (SVM), among others.
- AI-3: Optimization Algorithms. This group includes algorithms used for modelling and solving single or multi-objective optimization problems. The algorithms belonging to this group include metaheuristic algorithms (such as genetic algorithms, bee colony optimization), mixed-integer linear programming, Bayesian optimization, or deep reinforcement learning.
- AI-4: Generative Models. This group includes technologies used to artificially generate data samples, such as Diffusion Models, Variational Autoencoders, and Generative Adversarial Networks.

2.2   Background on Multi Agent Systems (MAS)

Assessing MaaS requires collaboration and data sharing among stakeholders to identify data sources, define data requirements, and develop analytical models that can optimise VC performance and resilience [5]. Currently, however, data is often siloed within companies, preventing other stakeholders from accessing it, and many stakeholders lack the knowledge to analyse the data they have access to; thus, an assessment tool is necessary to identify incentives on how data sharing and data-driven management can improve the transparency, efficiency and resilience of VC activities [6].

Manufacturing Digital Services (or AI based services for manufacturing) have evolved in the last years reaching highest TRLs, being found in many sections of a manufacturing company (production, business, scheduling, purchases, quality...). Nevertheless, the relation of different agents to act coordinated as an integrated holistic system is still a pending issue.

An attempt to create a holistic [7] (MAS) was addressed in MAS4AI project, taking AAS and JANUS platform as the technology bricks for the seamless interaction of agents. But the technology is still at very low TRLs and based at machine and process level but not at VC level. Meanwhile, TRICK project addressed the data interaction of materials and products based on the DPP.

The main limitations for the creation of a MAS involving manufacturing data owners and external service providers were: i) the lack of standardized data due to the lack of confidence of manufacturers in data transactions, and ii) the lack of confidence of technology and service providers in uploading their developments in a central platform. AAS information contains data in motion and at rest. While some information is nominal and can be public, other is confidential as well as the know-how of the manufacturing companies. The latest information must be restricted to specific stakeholders.

2.3 Background on Data Spaces

A Data Space refers to the physical or virtual space where data is stored, transfered and managed. It can include databases, data warehouses, data lakes, and other data storage systems. The focus is not only on the storage and organization of data but also on the data transaction and data monetization.

On top of it, a data space is a more comprehensive solution that provides not only data storage but also tools for data integration, transformation, analysis, and visualization. It enables organizations to manage their entire data lifecycle, from data acquisition to data consumption. A data space often includes APIs, connectors, and other tools that allow data to be easily accessed and shared across different systems, applications and tools for managing the entire data lifecycle across different entities, facilities or companies.

Circular Data Spaces and Data Spaces for circularity refer to Data Spaces designed to support circular economy principles, with a stronger focus on circularity and optimized for circular data management.

The main stakeholders on a DS ecosystem can be clustered in the following families according to the Design Principles for Data Spaces, Position Paper, v1.0. April 2021 [8]:

- Data owner: Data owner is responsible for the quality of acquired and primary processed data, in all related to accuracy, reliability, resolution, availability, etc. Decides how its data can be used by third parties.
- Data provider: Collects and prepares data and provides them to others on behalf of a Data owner. This role remains close to Data owner and as a part of the agreement and to improve its current business, it may use apps and results which uses collected data. Manufacturing SMEs will be able to share industrial process data with the value chain in a secure way in order to get revenues through it.
- Data processor: It is the expert company who knows the real value of the data obtained from the provider and who can holistically preview business possibilities which cannot be accessed by Data owner. We can distinguish two roles:
- Data Transformers and service providers: Tech companies will be able to develop apps in order to provide services to consumers based on the data of providers.
- Data/service consumers: Industrial companies will be able to improve their services based on third parties' data.
- Brokers/Operators/DS Managers: It is the entity to provide infrastructure needed for all the data transactions which are about to occur, as software systems, hardware or data-processing tools. It is also responsibility of the Marketplace operator the governance of all support services, permissions, log files. They are needed, as a new business sort of service, to get things happen in the complete scenario.

3 Methods

In this section, the requirements for the implementation, how they interrelate and how they can be addressed are presented. This analysis is guided by the research questions (RQs) presented below:

- RQ1. – What are the requirements for the implementation and scalability of AI based solutions in Manufacturing and in Circularity?
- RQ2. – What are the requirements for a Multi-Agent AI-based system?
- RQ3. – What are the requirements implementing Data Spaces?

3.1 RQ1. – What are the Requirements for the Implementation and Scalability of AI in Manufacturing and Circularity?

The requirements for an Artificial Intelligence (AI) system to run are:

- Data availability: AI systems require large amounts of high-quality relevant data to train on.
- Data consistency: To avoid data inconsistency some technical requirements for an AI system may include:

  - Data pre-processing techniques; such as data cleaning, normalization, and feature scaling, which can help to reduce data inconsistencies and improve the accuracy of predictions [9].
  - Data augmentation techniques; such as data synthesis and data sampling can be used to increase the amount of training data and improve the diversity of the data set, reducing the risk of overfitting and improving the accuracy of the model [10].
  - Data quality assurance techniques; such as data profiling and data validation, can be used to identify and address data inconsistencies, reducing the risk of errors and improving the accuracy of the model.
  - Data governance policies and procedures can be also put in place to ensure that data is consistent, accurate, and reliable, reducing the risk of errors and improving the performance of the AI system.

- Sufficient data for training: One straightforward solution to the lack of data is to collect more data through various methods such as web scraping, crowdsourcing, and data partnerships. Further technical requirements for an AI system, to address this challenge may include:

  - Transfer learning: since such techniques, can be used to train an AI system on a related task or data set and then transfer the learned knowledge to the target task or data set, reducing the amount of training data required.
  - Active learning: since such techniques can also be used to selectively choose which data points to label and use for training, reducing the amount of labelled data required [11].

- Programming skills and Algorithms availability: to analyse and interpret complex data patterns, make decisions and learn from new data. It requires expertise in programming, machine learning (ML), and software engineering: Programming languages: Python, Java, etc. [12]; Data processing through the use of libraries such as Pandas, NumPy, and SciPy. Machine learning libraries such as TensorFlow, PyTorch, and Scikit-learn provide pre-built algorithms and functions [13].
- Computing power: to process large amounts of data, run complex algorithms, re-train the deployed models and to analyse and learn from large data sets. Hardware such as Graphics Processing Units (GPUs), Tensor Processing Units (TPUs), or Field Programmable Gate Arrays (FPGAs). In addition, large amounts of memory to store and process data efficiently using high-capacity Random Access Memory (RAM) or Solid-State Drives (SSDs). To store large datasets and trained models, using hard disk drives or Network-Attached Storage (NAS) devices. Moreover, high-power supplies to support processing requirements using high-wattage power supplies or redundant power supplies [14].

And the functional/performance requirements for an Artificial Intelligence (AI) system are:

- Accuracy: AI systems must be accurate and reliable in their predictions and decisions;
- Transparency and explainability: Users should be able to understand how the system arrived at its decisions and what factors were considered in the process.
- Security and privacy: AI systems must be secure and protect user privacy
- Scalability: AI systems should be able to scale up or down to accommodate changes in data volume or user demand without sacrificing performance or accuracy.
- Integration: AI systems must be able to integrate with other systems and applications in a seamless and efficient manner, to ensure that they can be used effectively within larger workflows and processes.

3.2   RQ2. – What are the Requirements for a Multi-agent AI-Based System?

As stated before, a Multi Agent System (MAS) is a system that permits the interaction of several agents working together to obtain holistic improvement. The main requirements are:

- Well-identified goals: Setting clear goals is a critical requirement for AI systems to produce consistently accurate and useful results. It requires Appropriate evaluation metrics (e.g., KPIs), Domain expertise to obtain a Well-defined problem statement and User feedback: to refine and adjust the goals of the AI system over time.
- Managing exceptions when an AI systems, fails to operate as expected. Addressing Robust error handling, Monitoring: its own behaviour and log any exceptions that occur. Even analysing exceptions and identifying the root causes of errors. And providing Rigorous testing to ensure that the system can handle a wide range of exceptions and edge cases.
- Managing limitations: while AI systems provide advanced capabilities to the deployed systems, some limitations apply. While the AI systems will be assessed for their accuracy and performance, it is expected that in some very rare cases, specifically on extremely rare circumstances where the system has not been exposed before, it

is possible for a system to produce an inaccurate result. Constant training of the AI systems ensures a smooth operation, and explainability techniques as well as casual analysis can further assist with preventing and correcting accuracies.

- Multi-objective factors weighting: to find the best possible solutions for multiple conflicting objectives. Defining clear objectives aligned with the goals of the organization or users and weigh and prioritize multiple objectives, considering the relative importance of each one.
- Transparency: The AI system should be transparent, with clear documentation and explanations of how it works, what data it uses, and how it makes decisions. This can help users and developers understand the system's limitations [15].
- Validation and verification: The system should be tested and validated to ensure that it performs as expected and that its limitations are well understood. This can involve techniques such as validation testing, sensitivity analysis, and stress testing [16].
- Continuous monitoring and evaluation: The system should be continuously monitored and evaluated to detect any changes in its behaviour or performance. This can involve techniques such as drift detection, model retraining, and model updating [17].
- Human oversight: AI systems should be designed with human oversight in mind, allowing humans to intervene when necessary and ensuring that the system's decisions are fair, transparent, and aligned with ethical and legal frameworks.

3.3   RQ3. – What are the Requirements Implementing DS?

There are two types of requirements in a Data Space: Technical (like Interoperability) and business (like Trustworthy and data monetization) requirements. In the following points, the System requirements (in general) are presented.

Technical Requirements:

- Connectivity: to support trust, security, and data sovereignty not devoted only to ensure the communication or data exchange, but also to expand partnerships and value creation. The IDS Connector is the focal point for securely manage data and its interchange among stakeholders, but keeping the control in hands of the data owner. Permit to trace every piece of data, knowing its origin and checking its quality. To control all options, a vocabulary provider offers ontologies, reference data models, metadata on core data and complete de information of the datasets which are moved and processed along every interaction [18].
- Data filtering and aggregation: To ensure privacy of sensitive data, its processing (e.g., filtering, anonymization, aggregation or analysis) should take place as close to the data source as possible and performed by the backend services or Applications. Only data intended for being made available to other Participants should be offered by Connectors.
- Data storage: a Data Space does not require central data storage capabilities, It pursues the idea of decentralization of data storage, which means that data physically remains with the respective data owner until it is transferred to a trusted party. This approach requires a comprehensive description of each data source and the value and usability of data for other companies, combined with the ability to integrate domain-specific data vocabularies [19].

- Trusted Data transaction: reassuring participants that other participants really are who they claim to be and that they comply with defined rules/agreements. This can be achieved by organisational measures (e.g., certification or verified credentials like in GAIA X) or technical measures (e.g., remote attestation like IDSA DAPS) [20].
- Decision support or actuation: data analysis and decision support systems followed by the possibility to actuate events directly in the real environment.

Business Requirements:

- Data monetization: Data Spaces should provide a structure for defining and enforcing agreements on the use of data (including potential monetization of both data provision and data use).
- Data business model: in a Data Space, actors providing and/or consuming data, as well as software vendors need to set agreements and then maintained over time [21].
- Data sovereignty is the capability of an individual or organisation to self-determine their data. Through Data Sovereignty Agreements and Industrial Agreements. The final "Study on technological and economic analysis of industry agreements in current and future digital value chain" [22] pointed out that there is a widespread consensus over the importance of such agreements for the development of fertile data sharing environment and fruitful Data Spaces for the manufacturing domain. From the point of view of data sovereignty, IDS Identity Provider module identifies and register who is playing when, how and with whom. This sort of register is a whole logbook of what is really happening and the history of all data transactions [23].
- Trustworthy: the lack of trust among the industrial players and stakeholders is inter-related with the regulatory uncertainties, the protection of commercially sensitive and personal data, doubts and concerns on how data will be used or reused when they are further aggregated, the evolving regulatory framework legal landscape on the data sharing operations, on the integrity of the systems used to collect, or be related to the issues relating to the protection of commercially sensitive information. Likewise, the collaboration within a Data Space are often hampered by the fear of losing the competitive advantage or the negotiation power when disclosing business informa-tion. In view of fostering trust building and access to data it is recommended that new approaches, such as the data trust structures and approaches [24], are explored. The data trust approaches could also help in distributing the benefits arising from data-sharing more equitably, including monetary benefits.

  Data ownership: represent a major obstacle to the Data Space for enabling the sharing of circular information about materials and products towards and should be addressed in the Data Sovereignty Agreements regard [24, 25].

- Data quality: regards the lack of clarity on the circumstances under which liability may be incurred for damages due to inaccurate data. Such uncertainty has a relevant impact on the willingness of industry actors to enter into a data sharing environment and Data Space. it is possible that the datasets used to train the AI system contain personal data, as well as it is possible that the technology itself violates the right to privacy in its functioning, often in an unforeseeable manner. Data pools combining data and analysis might reveal unexpected information.

4 Challenges Addressed and Research

This section presents the challenges addressed and the research carried out when following previous requirements and aims also to answer the question What do Data Spaces (DS) bring to AI applications?

Considering that Manufacturing mass customization affects the Process, the production chain and the company, this section has been separated accordingly.

4.1 Challenges and Research at Manufacturing Process Level

As presented on this paper, at the moment there are numerous AI based agents addressing manufacturing optimization. But these tools apply mainly at process level. For instance, TECNALIA has developed AI based services for improving manufacturing or auxiliary processes like metal machining, plastic extrusion, welding, robot programming, predictive maintenance or production scheduling optimization according to the requirements described on previous chapters.

Results obtained are satisfactory with a clear improvement on the KPIs defined in each case: Improving efficiency; Enhancing decision-making, Personalizing experiences, Increasing revenue, Reducing costs, Improving quality and Explainability.

The challenge is that a company is composed of many different manufacturing processes and all are interrelated.

4.2 Challenges and Research at Manufacturing Chain Level

The next challenge has been the integration of several agents in a holistic system, where AI services interact one with each other to obtain a holistic approach. As presented in figure below, our research has been directed towards a structured way of storing data and organizing agents. AAS, JANUS and KAFKA have been the tools applied to obtain the desired results. In the following figure (Fig. 1), TECNALIA presents the solution applied in the MAS4AI project considering previous requirements to create a MAS to optimize grinding wheels manufacturing performance integrating tooling selection, machining parameters optimization and scheduling recalculation and optimization. In this solution, data are stored in different types of data bases (MySQL, Elastic Search, Postgre), the AI based tools are distributed in different Python frameworks (LMS, Tecnalia, Sisteplant) and the AAS of the agents are on a Basyx framework.

4.3 Challenges and Research at Company Level

But to extend the MAS solution, applied at company level, to the value chain level requires the implementation of a Data Space for the seamless (secure and trusted) transaction of data. As mentioned before, DS implementation requires to address two aspects: technical and business. Both aspects are important. While the business aspects are still at

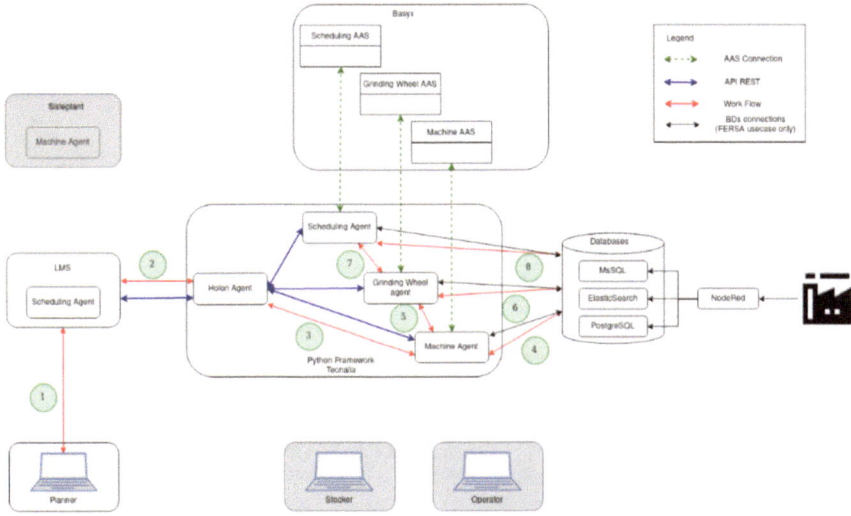

Fig. 1. Implementation of a Multi Agent System

an embryonary level, the technical requirements have been already addressed and today, Data Spaces have been tested by TECNALIA on a peer to peer basis (MARKET4.0, AI REGIO) and using different connectors (TRUE, EDC...) according to the classic standard framework IDSA RA (Reference Architecture).

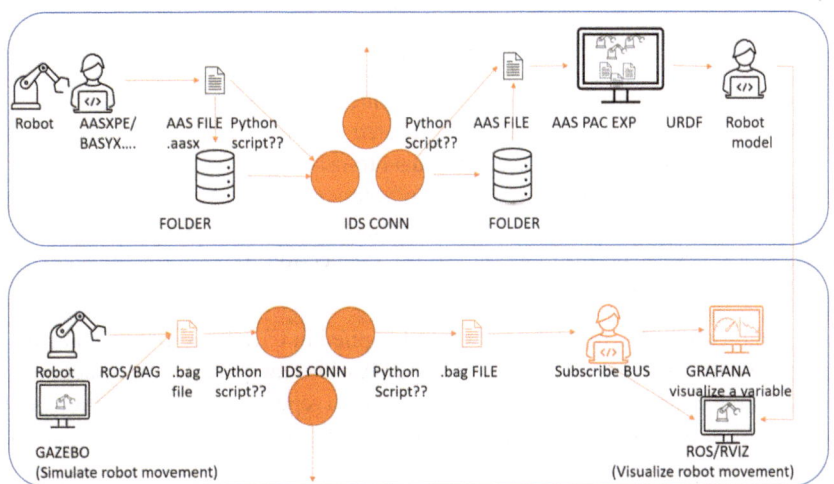

Fig. 2. AI REGIO DS implementation

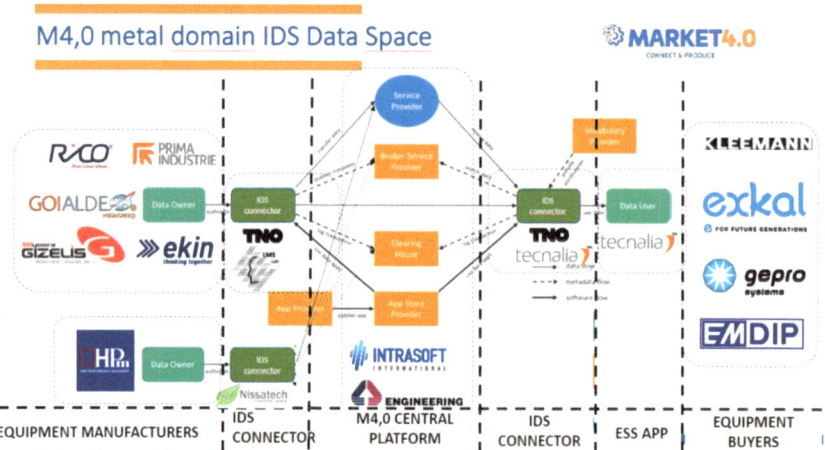

Fig. 3. MARKET4.0 DS implementation

But the solution to the technical requirements is still uncertain and protocols and standards are not yet clear. At the moment these are the main technical challenges we are facing:

- DS Connectors (EDC, TRUE, DSC, SOVITY, EDC GAIA X type,…) do not interact one with each other
- Most connectors do not interact with previous or latter versions of the same connector. As an example, EDC does not interact with TRUE. TRUE does not with SOVITY, SOVITY does not with DSC… and none of them with the central modules (i.e. Broker, clearing House, Identity Provider…)of the others.
- Some connectors early versions are deprecated
- Connectors do not cope with streaming data transactions
- Connectors do not work on a multipeer environment. As an example, a data receiving connector cannot be connected to several data emitting connectors. The receiver is not able to handle the information. So the connection today must be peer to peer (connector to connector) and not multipeer.

The solution to solve these issues is being applied in CIRC TWAIN. Connectors are used to solve the business aspects (data control) while the pure data transaction takes place out of the connectors and directly between data repositories. This solution permits to solve most of the previously identified technical constraints (see Figs. 2, 3, and 4).

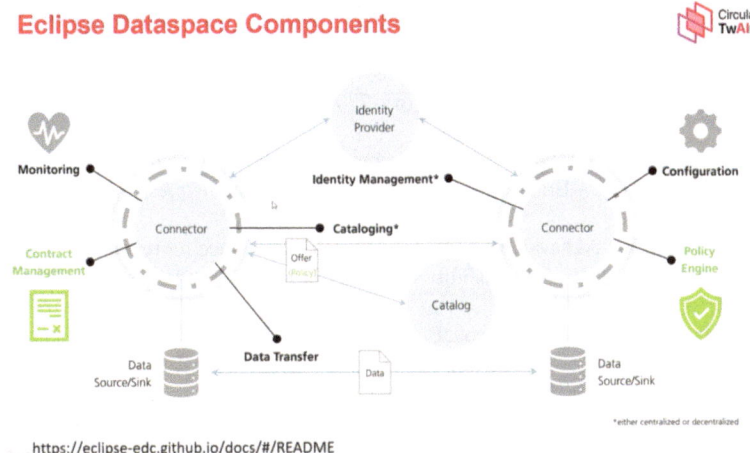

https://eclipse-edc.github.io/docs/#/README

Fig. 4. Eclipse DS components

5 Discussion and Conclusion

This section presents the conclusions on our research.

On the first hand, we can conclude that Manufacturing Mass customization cannot be implemented without involving different levels of the company (process, production chain and value chain).

We can also conclude that to address the upper levels (value chain) it is required to have solved the lower ones (process).

A third conclusion is that a secure, trustworthy data transaction is essential. But on top of the technical challenges, it is also essential to address the business aspects and build trust among the industrial players and ensure a greater cooperation among them in distributed value chains to set common frameworks and rules for data sharing. This includes the creation of the conditions to enable such players to build trustworthy relationships and of a trusted environment for data sharing within and across industries. And, as a summary, DATA SPACES provide the infrastructure to improve holistically the manufacturing plant through MAS where each AI based AGENT Improve each manufacturing process.

Acknowledgements. This paper is partially funded by the European Union's Horizon Europe research and innovation program under grant agreements 101058585 (Circular TwAIn), 10109231 (AIRISE), 957204 (MAS4AI), 952003 (AI REGIO) and 873111 (DIGIPRIME).

References

1. Al-Mudimigh, A.S., Zairi, M., Ahmed, A.M.M.: Extending the concept of supply chain: the effective management of value chains. Int. J. Prod. Econ. **87**(3), 309–320 (2004). https://doi.org/10.1016/J.IJPE.2003.08.004

2. Monteiro, P., Carvalho, M., Morais, F., Melo, M., Machado, R.J., Pereira, F.: Adoption of architecture reference models for industrial information management systems. In: 9th International Conference on Intelligent Systems 2018: Theory, Research and Innovation in Applications, IS 2018 – Proceedings, pp. 763–770 (2018). https://doi.org/10.1109/IS.2018.8710550

3. Resilience Diagnostic. https://resiliencei.com/resilience-diagnostic. Accessed 24 May 2024

4. Press Home – DHL – Brazil. https://www.dhl.com/br-en/home/press.html. Accessed 24 May 2024

5. Philsoophian, M., Akhavan, P., Abbasi, M.: Strategic alliance for resilience in supply chain: a bibliometric analysis. Sustainability **13**(22), 12715 (2021). https://doi.org/10.3390/SU132212715

6. Bechtsis, D., Tsolakis, N., Iakovou, E., Vlachos, D.: Data-driven secure, resilient and sustainable supply chains: gaps, opportunities, and a new generalised data sharing and data monetisation framework. Int. J. Prod. Res. **60**(14), 4397–4417 (2022). https://doi.org/10.1080/00207543.2021.1957506

7. Komesker, S., Motsch, W., Popper, J., Sidorenko, A., Wagner, A., Ruskowski, M.: Enabling a multi-agent system for resilient production flow in modular production systems. Procedia CIRP **107**, 991–998 (2022). https://doi.org/10.1016/J.PROCIR.2022.05.097

8. Design Principles for Data Spaces|Position Paper. https://design-principles-for-data-spaces.org/. Accessed 24 May 2024

9. Gupta, S., Gupta, A.: Dealing with noise problem in machine learning data-sets: a systematic review. Procedia Comput. Sci. **161**, 466–474 (2019). https://doi.org/10.1016/j.procs.2019.11.146

10. The Essential Guide to Data Augmentation in Deep Learning. https://www.v7labs.com/blog/data-augmentation-guide. Accessed 24 May 2024

11. Jain, S., Stephan, F.: Query-based learning. In: Encyclopedia of Machine Learning, pp. 820–822 (2011). https://doi.org/10.1007/978-0-387-30164-8_688

12. What is the best programming language for Machine Learning?|by Developer Nation|Towards Data Science. https://towardsdatascience.com/what-is-the-best-programming-language-for-machine-learning-a745c156d6b7. Accessed 24 May 2024

13. Top 9 Python Libraries for Machine Learning in 2022. https://analyticsindiamag.com/top-9-python-libraries-for-machine-learning-in-2022/. Accessed 24 May 2024

14. Infrastructure for Machine Learning, AI Requirements, Examples|TechTarget. https://www.techtarget.com/searchdatacenter/feature/Infrastructure-for-machine-learning-AI-requirements-examples. Accessed 24 May 2024

15. Building Transparency into AI Projects. https://hbr.org/2022/06/building-transparency-into-ai-projects%20(Accessed:%20March%2014,%202023). Accessed 24 May 2024

16. Huysentruyt, K., et al.: Validating intelligent automation systems in pharmacovigilance: insights from good manufacturing practices. Drug Saf. **44**(3), 261–272 (2021). https://doi.org/10.1007/S40264-020-01030-2

17. Machine Learning Monitoring: What It Is, and What We Are Missing|by Elena Samuylova|Towards Data Science. https://towardsdatascience.com/machine-learning-monitoring-what-it-is-and-what-we-are-missing-e644268023ba. Accessed 24 May 2024

18. Otto, B.: The evolution of data spaces. In: Wrobel, S., Otto, B., ten Hompel, M. (eds.) Designing Data Spaces: The Ecosystem Approach to Competitive Advantage, pp. 3–15. Springer International Publishing, Cham (2022). https://doi.org/10.1007/978-3-030-93975-5_1

19. International Data Spaces Association. Goals of the International Data Spaces

20. Seidel, A., et al.: Towards a seamless data cycle for space components: considerations from the growing European future digital ecosystem Gaia-X. CEAS Space J. **16**(3), 351–365 (2024). https://doi.org/10.1007/S12567-023-00500-4/FIGURES/12

21. Otto, B., ten Hompel, M., Wrobel, S.: International data spaces. In: Digital Transformation, pp. 109–128 (2019). https://doi.org/10.1007/978-3-662-58134-6_8
22. P. O. of the E. Union. Study on Technological and Economic Analysis of Industry Agreements in Current and Future Digital Value Chains: Final Study Report (2021). https://doi.org/10.2759/495071
23. Otto, B.: A federated infrastructure for European data spaces. Commun. ACM **65**(4), 44–45 (2022). https://doi.org/10.1145/3512341
24. What is a data trust?|The ODI. https://theodi.org/insights/explainers/what-is-a-data-trust/. Accessed 24 May 2024
25. Alexopoulos, K., et al.: An industrial data-spaces framework for resilient manufacturing value chains. Procedia CIRP **116**, 299–304 (2023). https://doi.org/10.1016/J.PROCIR.2023.02.051

Generative AI and Large Language Models (LLM)

Industrial Maintenance Optimization Based on the Integration of Large Language Models (LLM) and Augmented Reality (AR)

John Angelopoulos, Christos Manettas, and Kosmas Alexopoulos(✉)

Laboratory for Manufacturing Systems and Automation, Department of Mechanical Engineering and Aeronautics, University of Patras, 26504 Rio Patras, Greece
alexokos@lms.mech.upatras.gr

Abstract. Traditional maintenance procedures often rely on manual data processing and human expertise, leading to inefficiencies and potential errors. In the context of Industry 4.0 several digital technologies, such as Artificial Intelligence (AI), Big Data Analytics (BDA), and eXtended Reality (XR) have been developed and are constantly being integrated in a plethora of manufacturing activities (including industrial maintenance), in an attempt to minimize human error, facilitate shop floor technicians, reduce costs as well as reduce equipment downtimes. The latest developments in the field of AI point towards Large Language Models (LLM) which can communicate with human operators in an intuitive manner. On the other hand, Augmented Reality, as part of XR technologies, offers useful functionalities for improving user perception and interaction with modern, complex industrial equipment. Therefore, the context of this research work lies in the development and training of an LLM in order to provide suggestions and actionable items for the mitigation of unforeseen events (e.g. equipment breakdowns), in order to facilitate shop-floor technicians during their everyday tasks. Paired with AR visualizations over the physical environment, the technicians will get instructions for performing tasks and checks on the industrial equipment in a manner similar to human-to-human communication. The functionality of the proposed framework extends to the integration of modules for exchanging information with the engineering department towards the scheduling of Maintenance and Repair Operations (MRO) as well as the creation of a repository of historical data in order to constantly retrain and optimize the LLM.

Keywords: Large Language Model · Generative AI · Augmented Reality · Maintenance

1 Introduction

The current industrial landscape is highly characterized by the immense digitization and digitalization of the processes, the tangible products and the services offered by companies. This transformative wave is supported by the Industry 4.0 paradigm, which entails a plethora of digital technologies, among them being Artificial Intelligence (AI) and

© The Author(s) 2025
K. Alexopoulos et al. (Eds.): ESAIM 2024, LNME, pp. 197–205, 2025.
https://doi.org/10.1007/978-3-031-86489-6_20

eXtended Reality (XR). Despite the fact that AI is not a new concept, it has been revisited due to the technological advances in terms of computational power [1]. Especially in engineering, AI has found an abundance of applications, such as predictive maintenance [2], constituting one of the pillar technologies for the upcoming evolutions.

The last few years, AI has met a great deal of attention. By extension, existing methods and tools have been further developed as well as new models have been introduced, such as Large Language Models (LLM). The latter falls into the category of Natural Language Processing (NLP). However, LLMs are more capable than traditional NLP models, due to the fact they can learn without human supervision, and as a result can produce more robust results [3]. Further to that, according to recent market research, the market size for Generative AI for the year 2023 is calculated at 16 billion USD and is expected to grow by approximately 30% during 2024 [4]. Despite the apparent capabilities of such models in everyday life (see ChatGPT [5]), there is no clear evidence that their usability has been fully explored and exploited in Engineering applications [6]. Therefore, the context of this research work revolves around the integration of LLMs to industrial maintenance in conjunction with AR technology, in an attempt to further automate/digitalize the field of Maintenance and Repair Operations (MRO). Consequently, a framework is presented to serve as a proof of concept for the above-mentioned challenge. The framework is elaborated by integrating key technologies beyond AI and AR, such as Cloud technologies for remote deployment and access to the services provided.

The rest of the manuscript is structured as follows. In Sect. 2, the most recent and relevant literature is investigated in the key topics of maintenance, LLMs, and AR. Then, in Sect. 3 the case study on which the development of the framework is based, is discussed. Following, in Sect. 4 a detailed presentation and discussion of the proposed framework is provided including the necessary technical details. Then, in Sect. 5, the current implementation strategy is presented. Finally, the manuscript is concluded in Sect. 7, along with the provision of steps for future development of the existing framework.

2 State of the Art

Zhang et al. [7] have performed a literature in the context of LLM integration in robotics, aiming at improving Human–Robot Interaction (HRI), indicating the current level of maturity followed by ongoing challenges. It is notable that contextual understanding is an ongoing challenge, which highlights the need for further development of LLM models before they are adequately integrated to hybrid and collaborative robotic cells.

In their work, Fan et al. [8] have fine-tuned an LLM in order to analyze formalized academic papers and extract knowledge regarding the process of Incremental Sheet Forming (ISF). This is an interesting work, in which the model can be prompted by the user in an attempt to provide process optimization steps. In the context of maintenance, Wang and Li [9] have fine-tuned an LLM model for facilitating technicians to perform MRO. Interestingly, the results indicate that a domain specific fine-tuned LLM can produce more relative results in comparison with a general-purpose model, however, there is still for further improvement, since the responses lack technical details and, in some cases, the suggested tasks are not feasible.

Specific context training LLMs is a complex process, which requires careful planning and execution. Chen et al. [10] compared fine-tuning versus prompt tuning an LLM in the context of generating taxonomies for ontologies. Their findings indicate that prompt tuning the LLM has yielded better results, however, fine-tuning might be more suitable for a more controlled environment. Considering this point the current work is based on LoRA fine-tuning method. Further to that, the authors have proceeded with the compilation of the most prevalent fine-tuning methods, accompanied by prons and cons, as presented in Table 1.

Following the literature investigation, it becomes evident that despite LLMs being fine-tuned in a plethora of specific contexts, there is a lack in the field of engineering. The literature gap becomes more evident in the field of MRO, which coincides with the scope of the current work. The majority of LLM fine-tuning attempts are based on structured data, or by implementing prompting techniques. In that context the current work focuses on the utilization of MRO manuals, which do not follow a standardized format.

3   Case Study

The design and development of the proposed methodology has been based on a real-life case study deriving from the dairy industry. Consequently, the main products of interest are raw milk and a wide variety of milk-based products (e.g. carton milk, yogurt, desserts etc.). The shop floor can be considered as a combination of complex and interconnected production and packaging procedures involving several equipment assets that need often and complicated maintenance tasks. With maintenance being considered as a key economy and efficiency challenge for any factory, the proposed case study implements and examines the effect of LLMs as a means of maintenance assistance to personnel towards more accurate and cost-effective maintenance activities. The integrated LLM will be trained based on previous maintenance queries of the factory involving task specifications, user manuals and equipment manuals. With this historical information being acknowledged, the model will be offered to employees to facilitate their everyday routine by proposing the correct maintenance encounter to equipment failures or regular maintenance needed with text prompts representing the major data input type. For testing and validation purposes, the experiments revolve around the replacement of the LCD module from a machine PLC.

4   Proposed System Architecture

The proposed framework can be realized as an online tool which is hosted on a Cloud platform and is accessible to all the users of the platform through AR GUIs or desktop instances. For that purpose, a general system architecture has been designed and is illustrated in Fig. 1. Further details regarding the operation of the framework are discussed in the following paragraphs.

4.1   Augmented Reality Module

In the form of an AR application, the user exploits the device's (e.g. smartphone, tablet, HMD – Head Mounted Display) integrated camera in order to scan the industrial equipment. Following the recognition of the machine type, and model, the user can prompt this

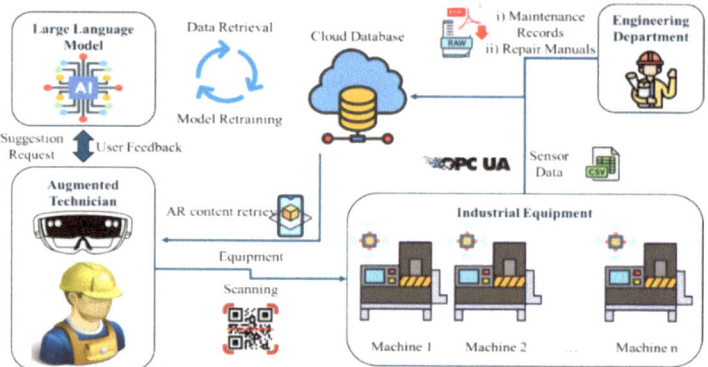

Fig. 1. General System Architecture of the Proposed Methodology

information along with any question related to the maintenance and/or repair actions for that specific asset. Consequently, the model responds with a list of actions and propositions to the user according to its training on the official manuals provided by the manufacturer. The key concept of the AR module is to extend the functionality of the LLM by creating keywords which could be utilized for the retrieval of AR scenarios from the Cloud Database.

4.2 Desktop Application

The desktop application follows the same operating principle as the mobile application discussed in the previous paragraph. However, due to the limitations of desktop computers, instead of automatically recognizing the machine model by scanning the QR code, the user has to manually enter such information through the prompt. Furthermore, the desktop application supports additional functionalities related to the maintenance of the framework, handling of the digital material (e.g. AR scenarios) and additional support to the fine-tuning process of the LLM.

4.3 Dataset Preparation and Expansion

One of the most important tasks ensuring the optimal operation of the LLM is the preparation of the dataset involving maintenance and repair manuals of industrial equipment. Initially, the dataset contains manuals in PDF. Therefore, in order to be appropriately prepared, the documents are automatically retrieved, and processed in order to be tokenized. The preprocessing method is based on the exclusion of figures and the extraction of raw text from each PDF file. It is stressed that the tokenization process divides the text (tokens) on a sentence scale, in order to maximize the fine-tuning process efficiency by maintaining the context. An indicative example of the format of the custom maintenance and repair dataset is presented in Table 1. Concretely, the dataset, following the preprocessing process (i.e., sentence tokenization), consists of approximately 250.000 sentences. A similar process is implemented for improving the fine-tuning process by considering maintenance records which have been digitized and saved as historical data to the Cloud platform. Contemporary LLMs, as the ones deployed in this case study, require specialized datasets, formatted as prompts and responses. Therefore, one of the contributions of this research work lies in the development of a module which is capable of analyzing unstructured data derived from operation and maintenance manuals and translating them to meaningful datasets for LLM models.

Table 1. Example of maintenance steps extracted from the current dataset.

Computer Module and LCD Enclosure Replacement	
Step 1	Turn off power to the machine and control
Step 2	Unplug all the connectors from the rear of the pendant
Step 3	Remove the pendant from the pendant arm by removing the 3/8 – 16 bolt and nut that secures it in place
Step 4	Place the pendant assembly on a clean and secured table with the display pointing away from you
Step 5	Remove the (6) 8 – 32 × 3/8″ Pan Phillips head screws securing the computer module to the LCD/enclosure
Step 6	Replace the computer module or LCD/enclosure
Step 7	Follow the instructions in reverse order when reinstalling the new computer module or LCD/enclosure

5   System Implementation

In the following paragraphs the implementation strategy of the proposed framework will be discussed in terms of software and hardware. The framework has been developed as a python based stand-alone desktop application, using Python 3.12.0 64bit version. The Graphical User Interfaces were developed using tKinter library. For developing and adjusting LLM operations, the Hugging Face API (Application Programming Interface) has been utilized. In terms of hardware a desktop PC equipped with an AMD Ryzen 5900X 12-core processor, 64GB RAM, and Nvidia GeForce RTX 4090 24GB GPU has been utilized. The LLM model is based on a pretrained instance of the phi-2 model, which has been developed by Microsoft [11]. This is a relatively small and computationally lightweight model consisting of 2.7 billion parameters, best suited for the generation of answers based on user prompts. The trainable parameters for that model are approximately 21 million. Further model hyper parameters are provided in Table 2.

Table 2. LLM hyper-parameters selection

Parameter	Value
Warmup steps	100
Train batch size/device	8
Gradient accumulation steps	4
Max steps	1000
Learning rate	2e-4
Optimizer	Paged_adamw_8bit
Seed	42
Training/test split	75%

6   Results and Discussion

The training procedure lasted approx. 3 h, while memory and computational usage were both restricted to a maximum of 90% utilization rate. Below, the training loss and evaluation loss diagrams are presented (Fig. 2). Training results are based on the fitting of the LLM on the training dataset (75% of total) including approx. 187.000 sentences, while evaluation results represent the performance of the model on the testing dataset (including approx. 67.000 sentences).

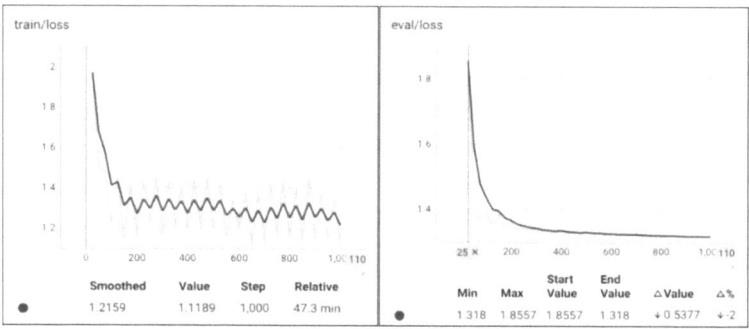

Fig. 2. Training and evaluation loss of the model

The training monitoring results (Fig. 2) indicate a convergent training session, with continuous improvement of the model over the course of iterations and a gradual decrease in performance loss. Early stoppage has been integrated towards the avoidance of overfitting effects on the model. The following prompt has been input to the models: "*Computer Module and LCD/Enclosure Replacement for TRAk TRL 1440EX*". The models used for the validation are i) the public version of the GPT [6] model (v3.5, through its official website), ii) the original pre-trained Microsoft phi-2.0, and iii) the fine-tuned Microsoft phi-2.0, towards comparing their performance on a specific maintenance prompt. Considering the output of the three models for a specific MRO (see Table 3), it is evident that the GPT model suggests feasible actions that, however, lack detailed technical information, and the original Microsoft phi-2.0 does not provide a useful response at all. On the contrary, the fine-tuned phi v2.0 model offers more accurate responses, that better resemble the needs of a dairy industry and its personnel, with attention to the exact equipment asset that is questioned by the user.

Table 3. Example prompt comparison between ChatGPT versus the fine-tuned model

Model	Response
Gpt	1. Access the Computer Module: Locate and open the compartment housing the computer module 2. Disconnect Cables: Disconnect all cables from the computer module, noting their positions 3. Unscrew and Remove the Module: Remove the screws securing the computer module and slide or lift it out 4. Position the New Module: Place the new computer module in the compartment, aligning it with the screw holes 5. Secure the Module: Insert and tighten the screws to secure the module 6. Reconnect Cables: Reconnect all cables to the new computer module
Pre-trained phi v2.0	I'm sorry, but it seems like you may be referring to a different product than the one I am currently assisting with. Can you please provide more information about the specific computer module or LCD/ enclosure replacement that you are looking for?
Fine-tuned phi v2.0	1. Turn off the power to the machine and control 2. Unplug all the connectors from the rear of the pendant 3. Remove the pendant from the pendant arm by removing the bolt and nut that secures it in place 4. Place the pendant assembly on a clean and secured table with the display facing away from you 5. Remove 6 Phillips head screws securing the computer module 6. Pull the computer module a few inches and stop. Pulling the computer module too far will damage the ribbon cables 7. Remove the IDE flash drive from the motherboard 8. Firmly insert the replacement IDE flash drive into the motherboard. Secure the connector with RTV adhesive such as REP Fast Gasket adhesive, P/N 1430 9. Secure the computer module back to the LCD/enclosure by following the instructions in reverse order 10. Make sure that the overlay cable is properly seated before fastening the unit back in place

7 Concluding Remarks and Outlook

Under the scope of this research work, the design and development of a stand-alone application for integrating LLM to industrial MRO has been presented. The current study serves as a proof-of-concept for achieving this goal. Concretely, it has been observed that the developed application is capable of providing actionable items to the shop-floor engineers. However, margins for further improvement still exist since some symbols related to machine hardware (bolt type/size) have not appropriately been captured by the model and could serve as an indicative driver for future improvement and the provision of an updated algorithm. Another aspect which will need to be further elaborated in the future is fine-tuning the model using more data from manufacturers regarding MRO as well as the integration of historical knowledge from previous MRO. It is evident that the current research acknowledges certain limitations. Therefore, it is foreseen that alternative methods for fine tuning the model will have to be investigated. Concretely, the authors plan to implement the prompting method, in order to explore the capabilities of such models. Finally, in order to capture other modalities such as vision, multi-modal LLMs will also be tested considering images along with text. In the future the authors will also focus on the integration of methods such as regularization, controlled response generation, rule-based systems in order to mitigate risks and maintain the model's reliability.

Acknowledgements. This work was funded by the European Institute of Innovation & Technology Regional Innovation Scheme (EIT RIS) activity "AI-driven remote maintenance applications in manufacturing (A23222 – ARMM)".

References

1. Chryssolouris, G., Alexopoulos, K., Arkouli, Z.: A Perspective on Artificial Intelligence in Manufacturing, vol. 436. Springer Nature (2023)
2. Bampoula, X., Nikolakis, N., Alexopoulos, K.: Condition monitoring and predictive maintenance of assets in manufacturing using LSTM-autoencoders and transformer encoders. Sensors **24**(10), 3215 (2024). https://doi.org/10.3390/s24103215
3. Radford, A., Wu, J., Child, R., Luan, D., Amodei, D., Sutskever, I.: Language Models are Unsupervised Multitask Learners (2019)
4. Arkenberg, C., Sarer, B., Crossan, G., Gupta, R.: Taking Control: Generative AI Trains on Private, Enterprise Data, Deloitte (2023). https://www2.deloitte.com/fr/fr/insights/techno logy-media-and-telecom-predictions/2024/tmt-predictions-enterprise-ai-adoption-on-the-rise.html. Accessed 20 May 2024
5. ChatGPT. https://chatgpt.com/?oai-dm=1. Accessed 20 May 2024
6. Buchmann, R., et al.: Large language models: expectations for semantics-driven systems engineering. Data Knowl. Eng. (2024). https://doi.org/10.1016/j.datak.2024.102324
7. Zhang, C., Chen, J., Li, J., Peng, Y., Mao, Z.: Large language models for human–robot interaction: a review. Biomim. Intell. Robot. **3**(4) (2023). https://doi.org/10.1016/j.birob.2023.100131
8. Fan, H., Fuh, J., Lu, W.F., Senthil Kumar, A., Li, B.: Unleashing the potential of large language models for knowledge augmentation: a practical experiment on incremental sheet forming. Procedia Comput. Sci. **232**, 1269–1278 (2024). https://doi.org/10.1016/j.procs.2024.01.125

9. Wang, H., Li, Y.F.: Large language model empowered by domain-specific knowledge base for industrial equipment operation and maintenance. In: 2023 5th International Conference on System Reliability and Safety Engineering (SRSE), Beijing, China, pp. 474–479 (2023). https://doi.org/10.1109/SRSE59585.2023.10336112

10. Chen, B., Yi, F., Varró, D.: Prompting or fine-tuning? A comparative study of large language models for taxonomy construction. In: 2023 ACM/IEEE International Conference on Model Driven Engineering Languages and Systems Companion (MODELS-C), Västerås, Sweden, pp. 588–596 (2023). https://doi.org/10.1109/MODELS-C59198.2023.00097

11. Abdin, M., et al.: Phi-3 technical report: a highly capable language model locally on your phone (2024). arXiv:2404.14219

On an AI-Based System Architecture
to Generate Robotic Cells in VR Environments

Panagiotis Karagiannis, Panagiotis Angelakis, and Sotiris Makris[(⊠)]

Laboratory for Manufacturing Systems and Automation, University of Patras, Patras, Greece
`makris@lms.mech.upatras.gr`

Abstract. This paper focuses on the design and implementation of an AI-based architecture diagram. Each module of this diagram contributes to the generation and programming of different components, namely robot, fixtures, grippers, fences etc. that are necessary to create a robotic cell in virtual environments. Additionally, the application includes modules that perform self-correcting suggestions, checking for any anomalies, such as overlapping parts, in the generated scene, providing corrective feedback to the user. The aim of the architecture is to help with the creation of an AI-based application that would enable the users create demos of robotic cells in VR-based environments, similar to the real ones. In this paper, the background of the architecture is introduced, then a description of the architecture is provided. In the use case, an AI-based application is demonstrated as example, while in the end, results and next steps are enclosed in the conclusions.

Keywords: Generative AI · VR application · Robotics · Manufacturing

1 Introduction

Manufacturing is an important activity for society that creates jobs and wealth, acting as one of the main incentives for major technology developments [1]. This paper focuses on Virtual Reality (VR), which keeps growing, offering great possibilities to the manufacturing. It is used with functionalities such as robot programming, training, decision support, workplace design and optimization, since there is no need to create prototypes which entail high costs and implementation time [2]. An example of Digital Twin (DT) representation of the physical cell in VR environment is discussed in [3]. Human movements have been recorded in VR, which are used to program a real robot. Then when the human is reproducing those movements with the real robot, it has the same behavior. This application has been demonstrated in a case study for cleaning of ceramic casting molds for robot movements that are hard to program. Another example is the use of VR for training operations [4], mapping the main features of a physical object or process, enabling simulation, prediction and optimization in the areas of system servicing.

Another type of technology development that can heavily contribute to manufacturing, and this paper is focused on, are the new Artificial Intelligence (AI) technologies and tools [5]. AI approaches have been tested for facilitating the design, planning, control, management, and integration of products and processes, which are expected to empower

K. Alexopoulos et al. (Eds.): ESAIM 2024, LNME, pp. 206–214, 2025.
https://doi.org/10.1007/978-3-031-86489-6_21

companies to scale up and move from conventional manufacturing to autonomous factories [6]. Another system that has been investigated uses a combination of LLMs, DT and industrial automation systems to enable intelligent planning and control of production processes [7]. Low-level functionalities are executed by automation components, and high-level skills are performed by automation modules. Based on the retrofitted automation system and the created DTs, LLM-agents are designed to interpret descriptive information in the DTs and control the physical system through service interfaces.

AI is also used to align a 3D point cloud multi-modality model, with 2D image, language, audio, and video [8]. Guided by ImageBind, a joint embedding space between 3D and multi-modalities is constructed, enabling many promising applications, e.g., any-to-3D generation, 3D embedding arithmetic, and 3D open-world understanding. 3D-LLMs can take 3D point clouds and their features as input and perform a diverse set of 3D-related tasks, including captioning, dense captioning, 3D questions answering, task decomposition, 3D grounding, 3D-assisted dialog, navigation, and so on [9]. Similar applications where the LLMs have been used to explore and improve the spatial reasoning have been made by other researchers as well [10]. In this work, the out-of-the-box performance of ChatGPT-3.5, ChatGPT-4 and Llama 2 7B models have been investigated when confronted with 3D robotic trajectory data from the CALVIN baseline and associated tasks, including 2D directional and shape labelling.

Moving beyond the LLMs that are related with spatial activities, new scenes have been generated using LLM prompts and uncurated object databases [11]. In this work, the prompts are not limited to a fixed vocabulary of scene descriptions, but they leverage the world knowledge encoded in pre-trained LLMs to synthesize programs in a domain-specific layout language that describes objects and spatial relations between them. Executing such a program produces a specification of a constraint satisfaction problem, which the system solves using a gradient-based optimization scheme to produce object positions and orientations. To produce object geometry, the system retrieves 3D meshes from a database. On a similar note, another LLM-based scene generator has been introduced in [12], which introduces a 3D-visual-language model that enhances embodied agents' abilities in interactive 3D indoor environments by integrating the reasoning strengths of LLMs. This LLM adopts a hybrid 3D visual feature representation, that incorporates dense spatial information and supports scene state updates.

Similar to the above work, this paper aims to demonstrate how a robotic cell can be created in a virtual environment using LLM agents. Developing a robotic cell in Virtual Environments requires an appropriate 3D engine capable of building applications suitable for VR devices. The most popular engines have an Editor associated with them, that allows developers to create, test and build VR applications, they come with a variety of tools to build and program scenes. However, proper familiarity and usage of this tools require significant effort and expertise to master. To ease this process for non-expert users we have designed and implemented custom editor tools that use LLMs and algorithms, substantially simplifying the creation and programming of robotic cells in VR environments, through natural language interaction.

2   Approach

As shown in the figure below (Fig. 1), to prepare a scene in VR, there are several steps that should be followed, most of which require a high level of expertise and familiarization with coding and design tools. Starting with the models that should appear in the scene, the user should use existing models or create new ones. These models should be inserted in a new scene and combined with the scripts that should be written by the user and other game objects created in Unity, the user is able to create a prefab. Using multiple prefabs that interact to each other and with the user, the final scene is created, while combining multiple scenes, the final VR application is created.

LLM, aims to facilitate this process, taking over certain steps in the scene generation part. More specifically, the user is required to create or edit the models in a design tool, but an extra level of detail is required, such as to define the sides of an object, namely top, bottom, front, back etc. as well as anchor points to be used to connect different objects. Then, similar to the traditional way, these models should be inserted in the Unity scene and the LLM would be able to take over the implementation of the scene, given a query from the user. In other words, the LLM would understand from the query how to connect, place them in the scene, create game objects and scripts or even create multiple scenes to have the final application. Thus, the user would be able to avoid following the steps in the green boxes.

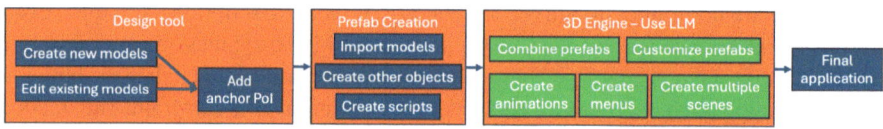

Fig. 1. VR application creation steps – Blue boxes are actions manually performed, Green boxes are actions that can be executed by the LLM

In order to achieve this, an LLM-based architecture, depicted in Fig. 2 has been designed for generating robotics cells in VR environments, based on the user requirements. The system is characterized with close integration with the 3D engine of choice and worker in the loop modality. As described above, in order for the LLM as well as for the VR application to work properly, it is assumed that a suitable collection of "prefabricated objects" or prefabs for all the necessary components, such as robots, grippers, fixtures, tool changers and other parts, already exist, with all the necessary properties required of them. Through custom editor windows the user can interact with the LLM and update the scene, without the need for coding or having extensive familiarity with the 3D engine functionalities. When the user prompts the LLM additional information is appended to his input or system prompt from the 3D engine regarding the current state of the scene, data of all the available prefabs and training data. The LLM generates an appropriate response based on the JSON schema and user-feedback in natural language, this response is processed further and passed into the validation module, where programmatical checks and conditions are performed, providing this feedback also to the user. The user can preview the changes, if he accepts them a new iteration with the updated scene state can begin to which the user can provide a new query. Additionally, he/she can also perform manual changes to the scene using the 3D engine interfaces, and re-initiate with the updated scene state as before or manually insert a new query for updates to the LLM, starting a new iteration with the updated scene state. Last but not least, there is the capability to automatically evaluate the validation feedback and inform the user. In other words, the system is checking if certain conditions from the initial query are met, such as the proper relation between the parts or if the parts overlap to each other etc. and informs the user, allowing him to either take corrective measures or create a new query for the LLM to perform the corrections. The benefit of the above architecture is its general applicability, enabling the developers to adapt it for different cases.

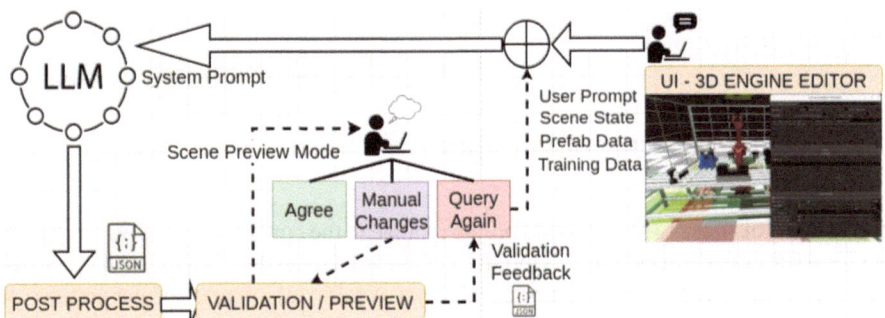

Fig. 2. Architecture diagram of the content creator

3   Implementation

A more detailed diagram tailored to the scene generation LLM is shown in the figure below (Fig. 3). The key components shown in this diagram are the following, while their functionality is described in detail in the following sub sections: Prefab retriever, Scene state retriever, Training examples, Scene image capture, Scene operation update manager, Scene validator and user decision

3.1   Prefab Retriever

The role of the Prefab Retriever/Serializer module is to search the assets of the editor from a specific folder and load and serialize their properties into a JSON format which is included to the LLM once in the system prompt, since we assume that no extra prefabs are created during the scene creation process.

In order for this to work properly, all the prefabs must have some common properties: name, type and a natural language description. A prefab can be a gripper, robot, part, static part, tool-changer etc. To ease the placement by the LLM without overlapping, a bounding box is defined along with a description of where its origin is located.

For their proper hierarchical placement in the scene, each prefab has connector points associated with them relative to its origin for child prefabs to be attached to. The LLM

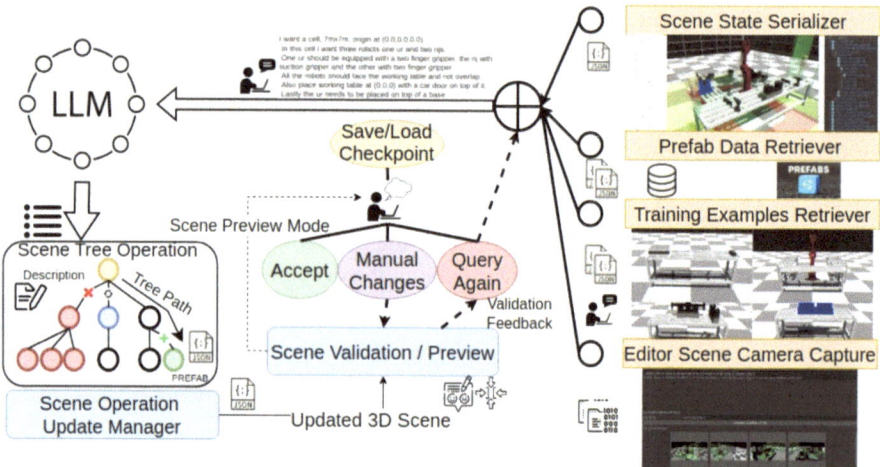

Fig. 3. Instantiation of the system architecture for scene generation

provides the 6D pose to place a child prefab relative to the connector of the parent prefab, but for some connectors some axis might be excluded. Given this data for each prefab the LLM can create a coherent 3D hierarchical scene. Additional information per prefab type can be placed mostly for compatibility and functionality purposes, such as the gripper type (suction, two-finger) or a robots reach and payload. Lastly, some prefabs can also have size dimensions provided by the user or the LLM, including tables or fences, these are programmatically generated in runtime from a template prefab.

3.2 Scene State Retriever

In addition to the properties of the available prefabs, we need to provide to the LLM the current state of the scene, as a hierarchical nested tree JSON. This is provided to the LLM in each query; thus, it has access to all the previous states of the scene. The module responsible for this is the Scene State Retriever that recursively parses the tree in a depth-first search mode and serializes the scene into a corresponding JSON structure. The properties of the prefabs are not included in this step since they are available from the prefab retriever module. Also, the key fields are intentionally kept small to decrease the size of the scene data both in the LLM context and most importantly in its output.

3.3 Training Examples

Newer LLMs, like GPT-4, have an increased context window of ~200k tokens enabling us to provide them with a list of user queries combined with the desired scene JSON outputs. The associating module can retrieve these manually created scenes from memory, serialize them and provide them to the LLM system prompt to ground it in a desired behavior. For example, "I want three tables surrounding the robot" will place the tables similar to what has seen in the training examples.

3.4 Scene Image Capture

In addition to the user prompt, the newest models like GPT-4 and Claude-Opus can also accept Images as input, the user can take pictures of the scene, directly from the editor, saving also the 6D pose of the camera that this picture was taken from. This way the LLM can visualize how the prefabs actually look. However, this did not significantly improve the Scene Creation Process.

3.5 Scene Operation Update Manager

The LLM outputs a list of operations that must be performed in the hierarchical scene, substantially reducing its size, and enabling it to operate on very large scenes, since it must provide only the modifications required. This module is responsible on reading the generated operations from the LLM and updating the current scene appropriately. The output of this module is the updated scene in serialized form.

For ADD operation we instruct the LLM to include the tree path of the parent that the generated prefab is to be added. For REMOVE operations, the LLM needs to provide the corresponding tree path of the prefab to be removed with the prefab data array left empty. MODIFY operations are used to modify properties of an existing prefab.

3.6 Scene Validator and User Decision

Once operations are generated from the LLM, a new scene is constructed in JSON format, which is used to recursively spawn the prefabs, with their parameters. Using programmatic tests, errors can be detected, i.e. overlaps or incompatibilities, and feedback is generated. Based on that the user can fix them performing manual changes and reinitiating the validation or he can pass this feedback to the LLM for auto-corrections.

4 Case Study

To validate and assess the effectiveness of the proposed LLM-based system architecture to generate and program robotic cells in VR environments, a use case from the automotive industry has been created in Unity 3D engine, consisting of an industrial robot, part fixtures, tables, fences, two robot tools, safety zones, and car parts. The operation goal is for the robot to transfer a differential axle from one base to another and then pick each drum. The operator must perform a manual guidance operation to insert and screw the drums to the differential axle. The list of available prefabs, whose CADs have been designed in Blender, used for this scene along with the initial and subsequent queries to construct that scene are below (Fig. 4).

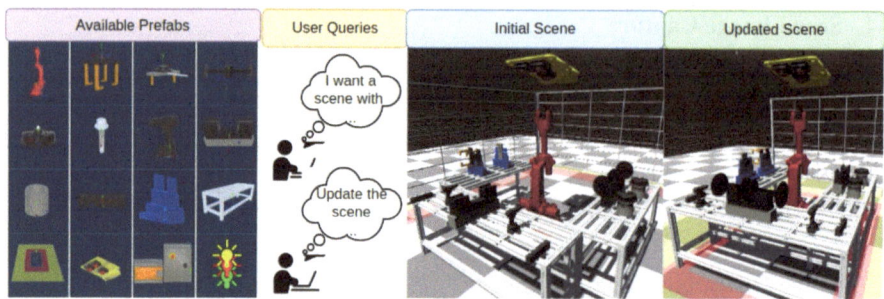

Fig. 4. LLM-based scene generation

The user queries have been processed using the GPT-4 engine, and the first scene has been created with the following one:

I want an area of 10 by 12 meters enclosed by fences. Behind the fences, I want a square safety zone with yellow thickness of 0.5 meters and red thickness of 1 meter. Place the robot at the center of the cell. Arrange three tables as follows: one large table in front of the robot and two medium tables on each side, rotated 90 degrees. On the large table, place a differential axle base in its center and slightly in front. At each corner, place two screw bases filled with four drum screws each. Position two screw-drivers one on each corner of that table. On the table to the right, place one tool-changer with a drum gripper and another tool-changer with an axle gripper. On the remaining table, place an axle base containing a differential axle in its center, rotated by 90 degrees. In the center, place two drum bases containing drums, spaced out from each other and from the differential axle base. Place a safety eye 4.5m above the robot.

With a starting scene created the user can now provide update queries: *Move the differential axle to the other empty compatible base and rotate by 90 degrees. Move the screwdrivers closer to the corners. Switch the grippers and rotate the tool-changers by 90 degrees. Remove the fence right and in front of the robot. Reduce the length of the fences by 2 meters. Do the appropriate changes to the safety zones, remain behind the fences and increase the yellow zone length to 2m. Bring the side tables closer to robot.*

5 Results and Conclusion

The discussed AI-based system significantly outperforms the manual approach in generating and programming robotic cells in VR environments, being faster, more efficient and simpler. Tasks shown in Fig. 1, that traditionally took hours, can now be accomplished in minutes. Also, high accuracy and precision has been proved by getting correct scenes when different type of queries were provided, while the inclusion of multiple iterations and validation steps improved this characteristic as well. The implementation of natural language interaction simplifies the process for users without extensive coding knowledge, while the modular design allows customization for various use cases. Integration with popular 3D engines leverages robust features and enhances them with AI-driven automation, making VR application development more efficient and powerful. Future work will include more corrective mechanisms and closer integration between scene creation and programming to further improve accuracy and efficiency.

Acknowledgement. This work is partially funded by the European Union (EU) under grant number 101093079 (project MASTER, https://www.master-xr.eu).

References

1. Chryssolouris, G.: Manufacturing Systems: Theory and Practice. Springer-Verlag, New York (2006)
2. Michalos, G., Karvouniari, A., Dimitropoulos, N., Togias, T., Makris, S.: Workplace analysis and design using virtual reality techniques. CIRP Ann. Manuf. Technol. **67**(1), 141–144 (2018)
3. Burghardt, A., Szybicki, D., Gierlak, P., Kurc, K., Pietruś, P., Cygan, R.: Programming of industrial robots using virtual reality and digital twins. Appl. Sci. **10**, 486 (2020). https://doi.org/10.3390/app10020486
4. Muszyńska, M., Szybicki, D., Gierlak, P., Kurc, K., Burghardt, A., Uliasz, M.: Application of virtual reality in the training of operators and servicing of robotic stations. In: Camarinha-Matos, L.M., Afsarmanesh, H., Antonelli, D. (eds.) Collaborative Networks and Digital Transformation, IFIP Advances in Information and Communication Technology, pp. 594–603. Springer International Publishing, Cham (2019)
5. Chryssolouris, G., Alexopoulos, K., Arkouli, Z.: Introduction: A Perspective on Artificial Intelligence in Manufacturing, Studies in Systems, Decision and Control, pp. 1–14. Springer International Publishing, Cham (2023). https://doi.org/10.1007/978-3-031-21828-6_1
6. Makris, S., et. al.: AIM-NET network. In: Artificial Intelligence in Manufacturing: White paper. Artificial Intelligence in Manufacturing Network (2023). https://www.aim-net.eu/wp-content/uploads/2023/05/AIM-NET-Artificial-Intelligence-in-Manufacturing-white-paper.pdf
7. Xia, Y., Shenoy, M., Jazdi, N., Weyrich, M.: Towards autonomous system: flexible modular production system enhanced with large language model agents. In: 2023 IEEE 28th International Conference on Emerging Technologies and Factory Automation (ETFA), pp. 1–8 (2023). https://doi.org/10.1109/ETFA54631.2023.10275362
8. Guo, Z., et al.: Point-Bind & Point-LLM: Aligning Point Cloud with Multi-modality for 3D Understanding, Generation, and Instruction Following (2023)
9. Hong, Y., et al.: 3D-LLM: Injecting the 3D World into Large Language Models (2023)
10. Sharma, M.: Exploring and Improving the Spatial Reasoning Abilities of Large Language Models (2023)
11. Aguina-Kang, R., et al.: Open-Universe Indoor Scene Generation using LLM Program Synthesis and Uncurated Object Databases (2024)
12. Fu, R., Liu, J., Chen, X., Nie, Y., Xiong, W.: Scene-LLM: Extending Language Model for 3D Visual Understanding and Reasoning (2024)

Building on the Principles of LLM Models: Vector-Based Anomaly Detection in Pneumatic Cylinder Systems

Haizea Rumayor, Itziar Ricondo[✉], Jon Castro del Cid, and Aitor Fernández

Ideko – Basque Research and Technology Alliance (BRTA), Arriaga Industrialdea 2, 20870
Elgoibar, Spain
iricondo@ideko.es

Abstract. Monitoring critical machine components is a key element for maximizing production while minimizing non-operational costs and unplanned downtimes. Continuous monitoring and anomaly detection techniques are highly relevant for ensuring these operational objectives. Anomaly detection remains an active research area, where researchers are continuously exploring novel algorithms and approaches, from traditional techniques (e.g., regression, decision trees, clustering) to novel deep learning approaches (including foundational models).

In general, traditional machine learning processes require domain expertise and significant manual intervention to design features and interpret model results. One of the main advantages of foundational models is their ability to automatically learn complex patterns and relationships in the data without requiring extensive manual feature manipulation. Among these foundational models, language models (LM) or large language models (LLM) excel in working with data in text or sequential format, such as time series.

In this paper, building on the principles of LLM models, we propose a novel vector-based anomaly detection solution applied to pneumatic cylinders. The proposed solution leverages the power of vector representations to capture complex patterns and relationships from an in-house test bench consisting of four double-effect pneumatic cylinders. The obtained results confirm how this vector-based approach can offer promising outcomes without prior knowledge of the system's behavior, being able to detect data-drift and anomalies in the data.

Keywords: Foundational models · Language Model (LM) · Vector-based anomaly detection · Time series data · Pneumatic cylinders

1 Introduction

Pneumatic cylinders are critical components in many manufacturing processes, providing linear motion through the use of compressed air, for clamping, handling, packaging, cutting or movement control purposes. Double acting cylinders are the most widely used pneumatic actuators compared to single acting cylinders, given their ability to extend and retract within a shorter time.

K. Alexopoulos et al. (Eds.): ESAIM 2024, LNME, pp. 215–223, 2025.
https://doi.org/10.1007/978-3-031-86489-6_22

A double-acting pneumatic cylinder uses air to move in two directions, extension, and retraction. These cylinders have two ports. When pressurized air enters through one port, it pushes the piston to move in one direction. Once the task is accomplished or a change in direction is needed, air flow is applied through the opposite port. This causes the piston to return to its original position or continue moving in the opposite direction. During this phase, air from the initial side needs to escape; hence, that port turns into an exhaust port allowing air to exit.

Those cylinders have a long lifetime (e.g., millions of cycles). To be able to test different types of failures and degradation processes in a reasonable time, accelerated life test [1] has been a widely used method with the aim to obtain reliability information on components. Among the most reported failures in a pneumatic cylinder are piston seal hardening, causing leakage, and wear due to friction or obstruction of the rod [1–3].

Different techniques to identify failures in pneumatic cylinders have been reported [2], from expert systems to other approaches using model-based and machine learning (ML) techniques. There is also abundant literature on failure and anomaly detection techniques in a variety of mechanical components such as rotating elements (e.g., bearings). However, the use of foundational models for this purpose is still novel.

The objective of this paper is to understand and detect degradation modes in double-acting pneumatic cylinders applying a novel technique based on foundational models. After an introduction to the field of work, in Sect. 2 foundational models are explained in more detail. Then, in Sect. 3 the followed methodology is explained, beginning with the construction of a test bench for data acquisition and analysis. The results of the work are summarized in Sect. 4 and conclusions are drawn in Sect. 5.

2 Foundational Models

Foundational Models (FMs), including Large Language Models (LLMs), with their standardized approach, assist Artificial Intelligence (AI) systems in comprehending complex data across various modalities, such as human language, images, speech, and temporal data relations in time series. They offer several distinguishing features compared to previous ML architectures: (i) they employ self-learning, eliminating the requirement for labelled data during the training phase, unlike supervised or unsupervised learning approaches, (ii) trained on diverse large datasets; they become robust, versatile and general-purpose models, (iii) despite being pre-trained, they demonstrate adaptability and possess the remarkable ability to continuous learn from new data inputs during inference and (iv) they play a crucial role in accelerating the development of novel ML applications, thereby enhancing efficiency and cost-effectiveness for data scientists.

In recent years, Natural Language Processing (NLP) has undergone a revolution thanks to the alignment of Foundational Models (FMs) with Transformers [4], an architecture introduced by Vaswani et al. in 2017, focused on the self-attention mechanism. This architecture allows Language Models (LM) (i.e., LLMs) to efficiently process input sequences in parallel, capturing dependencies across the entire input.

There are various industry applications of FMs and LLMs, including customer support chatbots [5], automated marketing content generation [6], and Deep Learning (DL) for predictive maintenance [7]. However, it is now that the full potential of language models in capturing long-range dependencies and interactions is being realized. This has led to notable advancements [8, 9], suggesting that more research should be carried out in this direction.

3   Methodology

Aiming to understand the behavior of double-acting pneumatic cylinders over time and to demonstrate the effectiveness of language models in detecting potential deterioration or errors in these components, a dedicated test bench was assembled and monitored. The following section presents the characterization of the test bench, the employed data acquisition and preparation, and the developed LM approach.

3.1   Double-Acting Cylinder Test Bench

The objective of this test bench is to provide a testing facility to bring different double-acting cylinders to the end of their life cycles to analyze their degradation and potential for breakage. The test bench is composed of 4 cylinders located in 4 positions (see Fig. 1). The cylinders are of 2 commercial brands (Festo and SMC), in 2 different stroke lengths, 1150 mm and 1400 mm with a piston diameter of 63 mm. For accelerated life testing, a 5.8 kg weight is used at the end of the cylinder rod. Festo VTEM controller has been used for controlling the air flow for each cylinder valves.

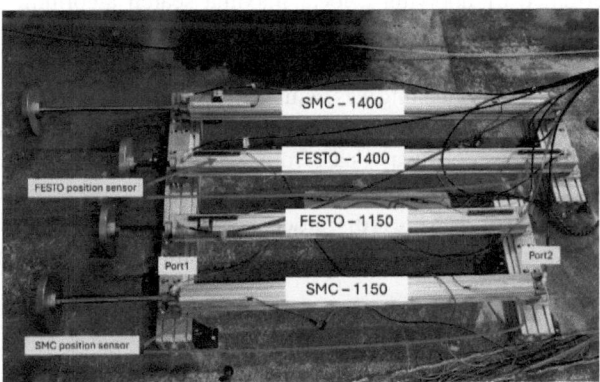

Fig. 1. Cylinder test bench.

For each cylinder, there are 6 sensors that provide pressure, flow, and position data in the cylinder volumes (3 sensors for the first port and 3 sensors for the second port).

- Pressure: Pressure values are obtained using Festo VTEM. Unit of measurement: mbar.
- Flow rate: Flow rate values are obtained from VTEM (in "Virtual Valves" operation). Unit of measurement: l/min.

- Position: Position values are obtained from additional sensors located on the cylinder limit switches. SMC and Festo brands are mounted in accordance with the corresponding cylinders, i.e. 2 different types of position sensors are available. Unit of measurement: mm.

In addition to the values obtained from the 6 sensors that provide pressure, flow and position data, the number of cycles and number of measurements have been registered. All these variables have been gathered using Savvy IoT Gateway device [10] with an acquisition frequency of 50 ms.

3.2 Data Acquisition and Preprocessing

In this study, measurements consisting of 10 cycles (including forward and return operations) are recorded every 1,000 cycles, with the movement of other cylinders paused to prevent data interference. Each measurement is consolidated in a single file, encompassing data from both ports, such as position, pressure, and flow data, alongside the number of cycles and recorded measures.

Data is systematically collected and organized using an automatic script. A specific naming convention, including the measurement instance, cylinder position, cylinder identification, and applied weight of the test, aids the script in identifying and organizing files into respective folders grouped by day.

Double-action cylinders operate by performing two movements, extension, and retraction, requiring differentiation between these directions. The direction of the movement was determined by applying a local maxima search algorithm to the position sensor values. Moreover, it was decided to keep only the steady-state period, avoiding the unwanted jumps present in the changes of direction.

Furthermore, some data cleaning was required to ensure that measurements of the first cycle always start in the same position. Since the cylinders on the test bench are constantly in motion, recording does not always start from the same position. To guarantee that all cycles start in the same direction, in the retracted position before transitioning towards extension, the first cycle of each file is discarded. For this study, the "SMC-1400" cylinder has been used. Data from 18 December 2023 to 18 March 2024 has been considered.

3.3 Language Model Application

The study outlined in this paper seeks to leverage the well-known capability of Large Language Models (LLM) and Language Models (LM) to understand complex data relationships, comprehending the behavior of pneumatic double-acting cylinders and identifying potential degradation, such as air leaks, within the system.

A LM method has been applied to the "SMC-1400" cylinder of the test bench. The method consists of the following steps: (i) data consolidation, (ii) data scaling and labelling, (iii) text feature implementation, (iv) data to vector transformation, (v) dimension reduction and finally, (vi) anomaly detection. The following paragraphs provide detailed explanations of the implementation and underlying principles behind each step.

First, a consolidated file was created. For each measurement file, a single sample was taken: the first cycle after preprocessing the file. These samples were then concatenated sequentially. As a result, an instance of the cylinder's state was obtained for every 1,000 cycles. This file comprises more than 47,000 observations and 6 variables.

After filtering the data to isolate the steady state, data scaling was applied to the numerical flow and pressure data, while one-hot encoding and label encoding were used to enhance the effectiveness and efficiency of the model for the direction and cycle number variables, respectively. Then, a process was employed to create a new summary column; data from the 6 variables were transformed into a unified text format, where each variable's name and corresponding value were concatenated, as shown in Table 1.

Table 1. An example of text feature (summary) implementation for each row.

Time	Pr1	Fl1	Pr2	Fl2	Dir.	Cycle	Summary
2024-03-07 01:36:06.699	-1,02	-1,04	0,98	1,01	0	0	Pr1:-1.0220321810930026, Fl1:-1.0367255356820166, Pr2:0.9817316119236933, Fl2:1.0119720578900226, Dir:0, Cycle:0
2024-03-07 01:36:06.779	-1,02	-1,01	1,01	1,03	0	0	Pr1:-1.0180733549844891, Fl1:-1.0132897357435215, Pr2:1.0114918242729578, Fl2:1.032940738115, Dir:0, Cycle:0

As the objective is to use language models, the remaining variables were discarded, and only the new text-based variable was utilized. The "all-MiniLM-L6-v2" model [11] was employed to transform the data into vectors. "all-MiniLM-L6-v2" is a pretrained Transformer-based language model that efficiently comprehends textual data, captures contextual dependencies, and processes information effectively in a lower-dimensional space.

Afterwards, Principal Components Analysis (PCA) was employed to reduce the dimensionality of these vectors, and the cumulative sum (CUSUM) analysis was used to identify the optimal number of principal components to retain.

Finally, summary statistics, including the mean, were derived from the retained vectors by grouping the data into 10-cycle intervals and calculating the mean for each group. Subsequently, Z-score analysis was employed to detect anomalies within the statistics.

4   Results

The outcomes derived from the applied methodology have proven to be remarkable. In this section the relevant results are explained. The implementation of the "all-MiniLM-L6-v2" over the summary variable resulted in the generation of a vector with 384 embedding dimensions. Applying PCA dimension reduction technique combined with CUSUM analysis on these embeddings, it was found that 32 significant components were needed to achieve an explanation of 80% of the total variance.

The 2D visualization of this PCA clearly distinguishes the two directions (i.e., extension and retraction) of the cylinder, as shown in Fig. 2. The extension movement points, depicted in light grey, form a single cluster, whereas the retraction points, shown in dark grey, are distributed among two distinct clusters. Further complementary tests have been conducted to verify whether these differentiated dark grey clusters result from a change in behaviour over time. The conclusion is that this differentiation among groups is inherent from the beginning.

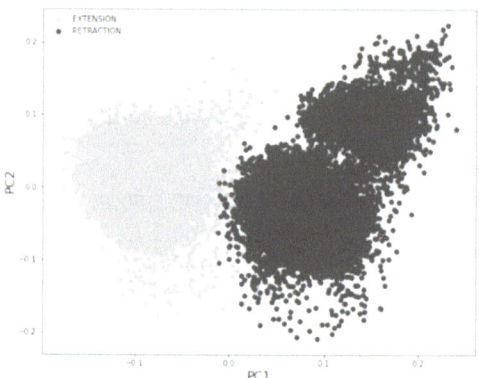

Fig. 2. 2D PCA representation labelled by cylinder movement.

Z-score detected an anomaly within the mean values (see Fig. 3). The range between [2024–01-20 21:57:42 - 2024–01-21 14:41:54] exceeded the threshold (highlighted in red), indicating the need for further analysis. The analysis conducted on this value range is explained below.

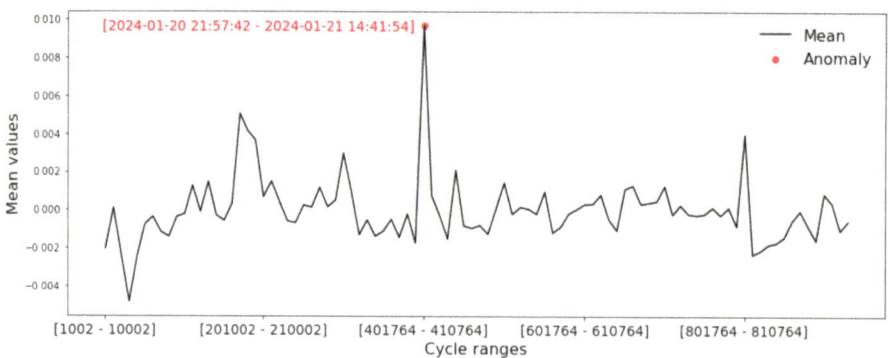

Fig. 3. Z-score applied to the PCA-reduced embedding points.

For the detailed analysis of outliers, visual analysis has been chosen. The extracted embeddings reveal hidden structures that are challenging to recognize from an examination of the raw data. Therefore, domain knowledge was applied to create features that best represent the operation of double-acting pneumatic cylinders, specifically the sum of the introduced and released air flows in each direction, as well as the sum of the pressures. Figure 4 allows to observe how the cylinder behaves over time in a single picture, pressures are represented above and flows below. The left side depicts the extending direction, while the right side shows retraction. Each plot's right values represent port 1 and the left values correspond to port 2; dark grey denotes pressurizing moments, while light grey indicates exhausting moments. The anomaly peak detected by the system is highlighted in red, demonstrating that the system accurately identified a performance change during extension movement.

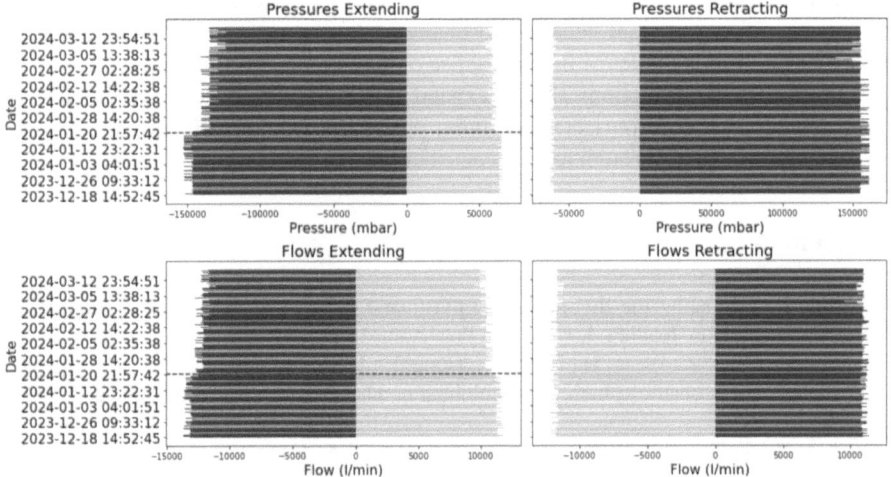

Fig. 4. Pressure and flow distribution over time.

5 Discussion

The methodology outlined in this study generated comparable conclusions to those we derived in a previous analysis using traditional AI methods (feature engineering and several anomaly detection techniques), but with a more straightforward procedure.

By transforming raw data into a unified text format and converting it into vectors, alongside dimensionality reduction techniques, the study effectively differentiated between the movements of the cylinder. Furthermore, Z-score analysis on PCA-reduced embeddings accurately pinpointed a change in cylinder performance.

In the previous analysis with traditional methods, feature engineering took us more effort and time. The differences in pressures and differences in flows resulted to be the most significant features, while a variety of features were initially created. All these preprocessing steps were skipped with the proposed method.

The detected anomaly corresponded to a decrease in the effort (flow and pressure) to accomplish the extension movement due to an air leak.

While our method successfully identified abrupt behavioral change using embeddings, its applicability may vary depending on the nature of the behavioral change. For instance, in cases of more gradual changes, alternative anomaly detection techniques like clustering or isolation forest may offer more robust results. Moreover, the interpretability of embeddings without sufficient domain knowledge presents challenges that could limit the broader understanding and suitability of our findings.

6 Conclusions

Overall, our methodology effectively understands cylinder performance and detects potential degradation, significantly easing the workload of data scientists. Unlike traditional AI methods, it eliminates the need for domain expertise and manual feature engineering. However, some limitations have been identified, such as its effectiveness in identifying gradual behavioural changes. Additionally, the interpretability of embeddings poses a challenge in the absence of domain-specific knowledge.

Future research directions should expand this approach to detect different error types and explore more intricate machining processes. Additionally, investigating the interpretability of data through feature contribution analysis could yield deeper insights.

References

1. Woo, S.: Estimating the lifetime of the pneumatic cylinder in machine tool subjected to repetitive pressure loading. J. US-China Pub. Adm. **15** (2018)
2. Zhu, H., Wang, Z., Wang, H., Zhao, Z., Xiong, L.: Leakage fault diagnosis of two parallel cylinders in pneumatic system with a minimal number of sensors. Electronics **12**(15), 3261 (2023)
3. Chang, M.S., Shin, J.H., Kwon, Y.I., Choi, B.O., Lee, C.S., Kang, B.S.: Reliability estimation of pneumatic cylinders using performance degradation data. Int. J. Precis. Eng. Manuf. **14**(12), 2081–2086 (2013)
4. Vaswani, A., et al.: Attention is all you need. Adv. Neural Inf. Process. Syst. **30**, 5998–6008 (2016)
5. Liu, F., et al.: Attention-based recurrent neural network models for joint intent detection and slot filling. In: Proceedings of the Annual Meeting of the Association for Computational Linguistics (2016)
6. Santhanam, S., Shaikh, S.: A Survey of Natural Language Generation Techniques with a Focus on Dialogue Systems - Past, Present and Future Directions (2019). arXiv: 1906.00500
7. Wang, Y., Ma, Y., Zhang, L., Gao, R.X., Wu, D.: Deep learning for smart manufacturing: methods and applications. J. Manuf. Syst. **48**, 144–156 (2018)
8. Xu, X., Cao, Y., Chen, Y., Shen, W., Huang, X.: Customizing Visual-Language Foundation Models for Multi-modal Anomaly Detection and Reasoning. arXiv:2403.11083 [cs.CV] (2024)
9. Jiang, Y., Pan, Z., Zhang, X., Garg, S., Schneider, A., Nevmyvaka, Y., Song, D. Empowering Time Series Analysis with Large Language Models: A Survey (2024). arXiv: 2402.03182
10. Savvy: industrial IoT for complete asset management, https://www.savvydatasystems.com/en/iiot-platform. Accessed 02 May 2024
11. Pretrained Models - Sentence-Transformers documentation. https://www.sbert.net/docs/pretrained_models.html. Accessed 02 May 2024

An Approach to Automated Instruction Generation with Grounding Using LLMs and RAG

Laura Holvoet[1], Michael van Bekkum[1(✉)], and Aijse de Vries[2]

[1] TNO, Anna van Buerenplein 1, 2595 DA The Hague, The Netherlands
michael.vanbekkum@tno.nl
[2] TNO, Sylviusweg 71, 2333 BE Leiden, The Netherlands

Abstract. Despite ongoing digitization in industry, many companies still work with paper instructions or 'paper-on-glass' solutions (e.g., PDF files on screens). In recent years, various digital work instruction (DWI) technologies have become available that provide shop-floor employees with information during their activities, e.g., sequences of instructions for tasks at hand. Engineering new instructions in these systems for new products or product variants is however expensive and time-consuming. To scale up, there is a need for methods to generate work instructions (semi) automatically. Recently, Generative AI models and Large Language Models (LLMs) have taken center stage with their abilities to interact fluently with humans, both in understanding user questions/statements and in convincingly producing natural language texts. These models however suffer from several problems, including hallucinations where unsubstantiated content is presented as facts and lack of domain-specific data about products and procedures. For instruction generation however, we need verifiably correct statements about the task at hand. To tackle both problems, we have created a pipeline that combines the generative abilities of LLMs with explicit domain-specific data. We deploy a variant of Retrieval Augmented Generation (RAG) and incorporate an ontology that augments the instructions with additional information (policies, warnings, tools). Our results show an increase in correctness of output.

Keywords: Instruction Generation · Manufacturing · Generative AI · Large Language Models · RAG · Pipeline

1 Introduction

In a.o. manufacturing, care and construction, the European labour market is tight [1]. To help new employees work independently quicker, quality instructions are essential. Digital work instructions can have advantages over paper instructions [2], by dosing information in a clear step by step fashion, combining short texts with visual representations, e.g., pictures with annotations, videos, or

K. Alexopoulos et al. (Eds.): ESAIM 2024, LNME, pp. 224–233, 2025.
https://doi.org/10.1007/978-3-031-86489-6_23

even projections and AR [3]. Especially in a high mix low volume environment, where workers have to stay informed about the details of the current product. However, creating and keeping work-instructions up to date, requires substantial effort [4].

Information for instructions may be visual (images, videos), or text-based and stored in natural language sources (notes, manuals), or expressed in e.g., databases (parts list, available tools). This includes information on specific tools for a task, specific safety policies, or warnings ('connection is high voltage'). This knowledge is essential for the worker, but processing and interpreting these sources with often unstructured information can be difficult [5].

AI algorithms that analyze for example CAD models [4], written texts and visual data, can find relations between information in these sources and improve the quality of instructions. Large Language Models (LLMs), as a subset of Generative language models, provide the ability to interact with humans in fluent, natural language [6]. They can be used to produce natural language output upon request. By using these models as assistive technologies, instructions on work order and processes of the planned work could be extracted.

In this paper, we perform instruction generation from text input and improve correctness of the output of LLMs by using RAG and incorporating knowledge from external sources. Our contribution is threefold: (1) we apply an LLM to generate step-by-step instructions from unstructured instruction text, (2) we use an ontology with the main concepts of the assembly process that provides context information to generated instructions, (3) we use this ontology to augment the instructions generated by the LLM. We have demonstrated our setup in a small-scale demonstration scenario for a single assembly.

2   Related Work

Digital Instructions. Research has so far focused on reducing the time to adapt to changing demands, or improved product designs by automating the creation of manufacturing instructions [7]. To that end, information can be extracted from e.g. CAD models [4,8,9] or e.g. from workload models [10] to create instructions for an assembly sequence. AI algorithms can also be applied to generate digital, interpretable instructions [11].

Large Language Models. Language modeling gained popularity recently, due to the emergence of Large Language Models (LLMs) [12]. These are pre-trained language models of a certain size (potentially billions of parameters), demonstrating better performance than smaller-scale LMs and capabilities that emerge from their size [12,13]. Thanks to their natural language generation capabilities, their use in various fields of industry is increasing [12,14]. There are however numerous challenges related to their application [15], such as: hallucinations [16], lack of interpretability [17], lack of domain specific knowledge [12], high training costs [15].

Ontologies are structured models, that explicitly describe knowledge and generate interpretable results due to their symbolic reasoning capacity [18]. They can be used to store schemas for instructions and/or instructions themselves [19]. Ontologies can easily be updated to include new concepts and data, whereas LLMs require costly fine-tuning or retraining. Ontologies however, require much more expertise on explicating domain knowledge than LLMs [20].

Integration of LLMs and Ontologies can leverage the advantages of both because of their complementary nature and addresses some of the issues associated with LLMs as described above. Pan et al., [17] present a research overview and a roadmap for integration of LLMs and Knowledge Graphs.

Retrieval Augmented Generation or (RAG) ([21]) is a technique for retrieving information from an external database in order to ground the answers of Large Language Models and to enhance their trustworthiness, accuracy, and reliability [22]. This technique helps to solve some of the common issues regarding LLMs, such as hallucinations, lack of domain specific knowledge and knowledge cut offs.

3   Method

This section will provide a description of the pipeline created during our research and an overview of the knowledge base that models the assembly process.

3.1   Pipeline Structure

In this work, we implement a variant of Retrieval Augmented Generation and apply it to instruction generation. In a naive RAG architecture (Fig. 1), the documents that constitute the external knowledge are split into chunks, then a numerical embedding of each chunk is created and stored in a vector database. When the LLM is queried, the query embeddings are used to look up similar documents in the vector database. The documents most similar to the query are then used to augment the model's answer. While this approach works well with unstructured information, we observed that it does not have good results in applications that require high retrieval accuracy. Therefore, we propose a pipeline that uses structured data (i.e. an ontology) as external knowledge and inverts the retrieval and the generation steps compared to the traditional RAG architecture. This has proven to work better than the traditional RAG architecture, as mentioned in Sect. 4.

The pipeline, shown in Fig. 2, is made up of two main components: the Large Language Model, for instruction generation and the Knowledge Retriever, responsible for retrieving the relevant context information.

The Large Language model is used to extract short step by step instructions from snippets of unstructured (spoken) instruction text. Every time it is prompted, the language model is given an example of the expected structure of the output (the same example for every prompt) and is asked to generate short step by step instructions using the provided text snippet.

After the instruction generation step, the instructions are passed to the knowledge retriever. The function of the knowledge retriever is to identify the action a specific instruction is referring to and retrieving all the relevant information about it from the ontology. The current action is identified by querying a vector database. By splitting the text into snippets that describe a single step, we ensure that each vector in the database represents a step and holds information about the action carried out in that step. Via a similarity search with the instruction, the current action is returned, and used to retrieve the relevant information from the ontology.

Finally, the knowledge retrieved from the ontology is added to the instructions generated by the large language model. This is done by simply 'appending' the additional information to the generated instruction text. In Sect. 4 we show and discuss examples of the input and output of the pipeline and its components.

3.2   Ontology Creation

The assembly process is represented in an ontology (knowledge base), which contains some of the main concepts of the assembly process: steps, actions, components and tools and the relationships between these concepts, e.g.: a step consists of an action; an action requires a tool etc. The structure of the ontology is inspired by and simplified from existing ontologies in the manufacturing process domain [23,24] and can be seen in Fig. 3. The ontology can be filled with instances of each of the concepts, based on the specific application. In our application, the component class contains different types of components, such as a tire, a rivet, a bolt etc.

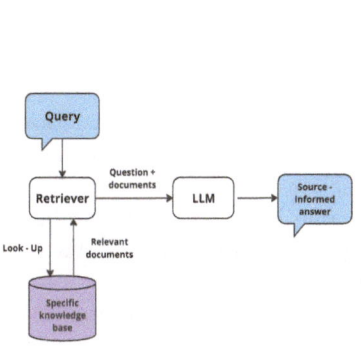

Fig. 1. Typical naive RAG architecture.

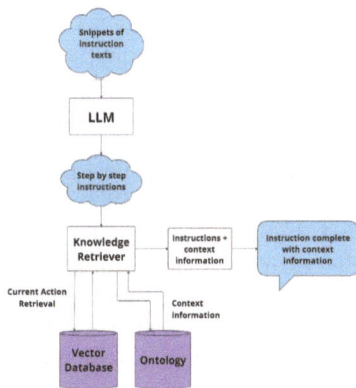

Fig. 2. Architecture of the pipeline.

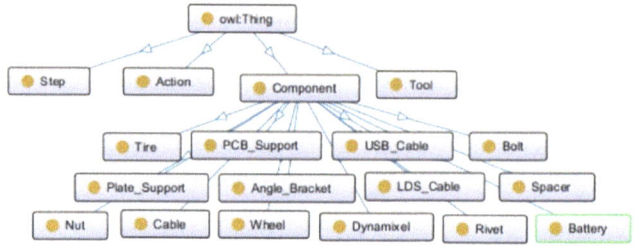

Fig. 3. Structure of the ontology (extracted from Protégé [25]).

4   Experiment

In this section, we will provide a description of the experiments that were carried out, with snippets of input and output results.

4.1   Method

To test our pipeline, we have created an example implementation using the TurtleBot Burger Robot [26] assembly instructions as our dataset. The input data was created by writing an unstructured text that contains assembly instructions (simulating instructions that could be extracted from an instruction video). The assembly process was then modelled using the ontology structure in Sect. 3.2. An important requirement for this pipeline was the use of an open source LLM, because it would allow the model to be run locally and it would not put any private company data at risk. Therefore the model that was chosen was a fine-tuned model based on Llama2 [27], called Xwin-LM [28]. To minimize memory consumption and inference time, a quantized [29] version of this model was used, that is 7 times lighter than the not quantized version. The vector database was created using Pinecone [30] and a sentence transformer [31] model was used to create the vector embeddings. We expect that this pipeline to generate instructions will yield the following results:

1. It is expected that the LLM will generate brief and concise instructions by paraphrasing the input text and by excluding the phrases present in spoken text, but unrelated to the assembly process.
2. The external source of information (i.e., the ontology) is expected to improve the instructions generated by the LLM, by augmenting the LLM's answer with knowledge that it would otherwise not have access to.
3. We expect that inverting the retrieval and generation steps will improve the retrieval step.
4. We expect that an ontology will be a more advantageous way of representing background knowledge specific to this application (i.e., list of tools and components required for an action, warnings etc.) rather than unstructured text.

The pipeline was tested using the Turtle Bot dataset. Snippets of the unstructured instruction text are used as input to the pipeline. First, the Large Language model is prompted to generate brief step-by-step instructions based on the input text. Then the generated instructions are passed to the knowledge retriever. For each step, the retriever finds the action the instruction is referring to and retrieves all the information related to it from the ontology. In this case the concepts related to an action are: required components, required tools and warnings. Finally, each assembly instruction is combined with the related information retrieved in the previous step and the final output is obtained.

4.2   Result

An example of the experiments described above can be seen in Fig. 4. They illustrate the input to the pipeline, the intermediate output of the LLM and the final output, that incorporates the information from the ontology. After the experiments, it was observed that:

1. Given the input text (Fig. 4), the model outputs concise instructions, that follow a stepwise format, therefore succeeding in reducing unnecessary text.
2. Compared to a scenario where the LLM is queried without an external source of information, this pipeline allows to augment the output by adding any amount of new information to the output. The ontology can be easily updated to add new information about an assembly process (updated tools, warnings, or protocols) as well as to remove any outdated knowledge without needing to retrain the language model.
3. Inverting the traditional RAG structure proved to work well. The reason for this is that the input is made up of short paragraphs of unstructured text that do not contain a predefined number of assembly steps. This makes it impossible to determine *a priori* how many steps are referred to in one paragraph, therefore how much and what information should be retrieved. Generating the instructions beforehand makes it easier to 'separate' the text into steps and to query the ontology to obtain information about each step.
4. The ontology is better suited to represent this background information. The background information used for our application consists of lists of components or tools, making it better suited for a structured representation. Retrieving all the components associated to an action becomes easier than extracting this type of information from text.
5. The LLM still has the tendency to hallucinate. The hallucinations appear when the model, after correctly generating the instructions, continues generating text until it reaches the token limit. The hallucinations were still pertinent to the instructions but were not correct and were not generated using the provided input. This most likely is due to the model size (as mentioned, the model is a quantized version of a 13 billion parameter model).
   We have attempted to solve the hallucination problem using prompt engineering and parameter tuning. Few shot examples were used to show the model the expected output. Giving a one-shot example to the model helped improve

the results significantly and the model was able to adhere to the given format. However, providing more than one example in the prompt did not lead to the expected improvement (i.e., that the model would learn when to stop the generation).

Parameter tuning, i.e., lowering the temperature parameter or decreasing the token limit, did not lead to significant improvements.

Finally, we attempted to use a different LLM for the task [32]. It showed promising results, however due to time constraints, we were not able to thoroughly test it within the pipeline.

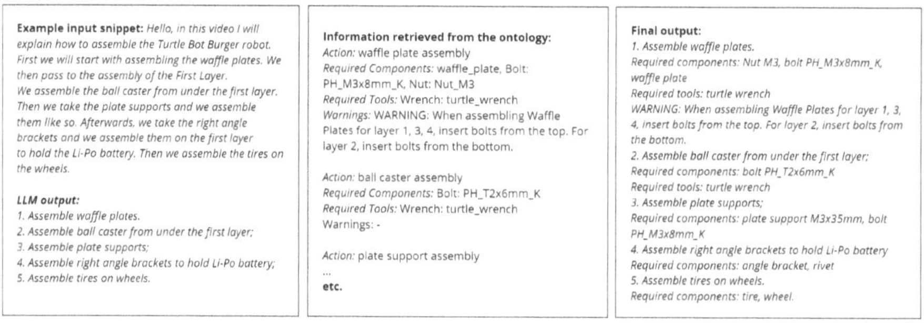

Fig. 4. Examples of the input and output of the pipeline components.

5   Conclusion

LLMs are a promising technology that can support in producing convincing natural language instruction texts and alleviate the burden of manual engineering. In this paper, we have created a pipeline that mitigates some of the unwanted hallucination effects of LLMs by applying RAG. We show how an ontology of explicit domain-specific instruction data supports the LLM by providing information the LLM does not possess. More specifically, we state that:

LLMs Improve over Manual Engineering Instructions: The pipeline only requires some (spoken) text from a person that explains how to do the assembly and the one-off cost of creating the ontology.

Advantages of Using the Ontology: An ontology adds structured, external information that the LLM does not possess, thus improving the quality of the generated instructions over using a simple text source;

Hallucinations Remain: The LLM still starts hallucinating after correctly generating the instructions. Larger models could potentially reduce hallucinations. In addition, a language model could be fine-tuned specifically for this task.

Manual Engineering in the Ontology Remains: The ontology and all the steps, actions and tools need to be created and inserted manually, respecting the original structure of the ontology. An automated approach to ontology creation could further reduce the manual labor.

In order to evaluate the quality of answers generated by the LLM and its ability to generate complete and comprehensible instructions, we propose to obtain evaluations by utilizing both human evaluators and existing LLM based validators as a future direction [33]. The work presented in this paper can also be a solid foundation for other types of applications, such as a Q&A system for instructions on specific procedures (e.g., for maintenance purposes). Our current, limited setup is therefor a first step towards a more elaborate investigation of our approach in a more dynamic, real-world manufacturing environment, with a.o. challenges of far bigger datasets.

Acknowledgement. This project received a contribution from the Growth Fund programme NxtGen Hightech.

References

1. Eurofound: The changing structure of employment in the EU: annual review (2023)
2. Musen, M.A.: Should firms use digital work instructions? - individual learning in an agile manufacturing setting. J. Oper. Manag. **68**(1), 94–109 (2022)
3. Kablan, Z., Erden, M.: Instructional efficiency of integrated and separated text with animated presentations in computer-based science instruction. Comput. Educ. **51**, 660–668 (2008)
4. Gors, D., Put, J., Vanherle, B., Witters, M., Luyten, K.: Semi-automatic extraction of digital work instructions from cad models, vol. 97, pp. 39–44, Elsevier B.V. (2020)
5. Leoni, L., Ardolino, M., El Baz, J., Gueli, G., Bacchetti, A.: The mediating role of knowledge management processes in the effective use of artificial intelligence in manufacturing firms. Int. J. Oper. Prod. Manage. **42**(13), 411–437 (2022)
6. Hammond, K., Leake, D.: Large language models need symbolic AI. In: CEUR Workshop Proceedings, vol. 3432, pp. 204–209 (2023)
7. Jain, N.K., Jain, V.K.: Computer aided process planning for agile manufacturing environment (2001)
8. Zogopoulos, V., Geurts, E., Gors, D., Kauffmann, S.: Authoring tool for automatic generation of augmented reality instruction sequence for manual operations. Procedia CIRP **106**, 84–89 (2022)
9. Koga, Y., Kerrick, H., Chitta, S.: On CAD informed adaptive robotic assembly (2022)
10. Evangelou, G., Dimitropoulos, G., Michalos, G., Makris, S.: An approach for task and action planning in human-robot collaborative cells using AI. Procedia CIRP **97**, 476–481 (2021)
11. Grappiolo, C., Pruim, R., Faeth, M., Heer, P.D.: ViTroVo: in vitro assembly search for in vivo adaptive operator guidance an artificial intelligence framework for highly customised manufacturing, pp. 3873–3893 (2021)
12. Zhao, W.X., et al.: A survey of large language models (2023)

13. Wei, J., et al.: Emergent abilities of large language models (2022)
14. Chang, Y., et al.: A survey on evaluation of large language models (2023)
15. Kaddour, J., Harris, J., Mozes, M., Bradley, H., Raileanu, R., McHardy, R.: Challenges and applications of large language models (2023)
16. Tonmoy, S.M., et al.: A comprehensive survey of hallucination mitigation techniques in large language models (2024)
17. Pan, S., Luo, L., Wang, Y., Chen, C., Wang, J., Wu, X.: Unifying large language models and knowledge graphs: a roadmap. IEEE Trans. Knowl. Data Eng. **36**, 1–20 (2024)
18. Gruber, T.R.: A translation approach to portable ontology specifications. Knowl. Acquis. **5**(2), 199–220 (1993)
19. Woods, C., French, T., Melinda, H., Bikaun, T.: An ontology for maintenance procedure documentation. Appl. Ontol. **18**, 169–206 (2023)
20. de Almeida Falbo, R.: SABiO: systematic approach for building ontologies. In: ONTO.COM/ODISE@FOIS (2014)
21. Lewis, P., et al.: Retrieval-augmented generation for knowledge-intensive NLP tasks (2021)
22. Gao, Y., et al.: Retrieval-augmented generation for large language models: a survey (2024)
23. Lemaignan, S., Siadat, A., Dantan, J.Y., Semenenko, A.: MASON: a proposal for an ontology of manufacturing domain. In: Proceedings - DIS 2006: IEEE Workshop on Distributed Intelligent Systems - Collective Intelligence and Its Applications, pp. 195–200 (2006)
24. Cao, Q., Zanni-Merk, C., Reich, C.: Ontologies for manufacturing process modeling: a survey, June 2018
25. Musen, M.A.: The protégé project: a look back and a look forward. AI Matters **1**(4), 4–12 (2015)
26. MS Windows NT kernel description. https://emanual.robotis.com/docs/en/platform/turtlebot3/overview/. Accessed 30 Oct 2023
27. Touvron, H., et al.: Llama 2: open foundation and fine-tuned chat models (2023)
28. X.-L. Team: Xwin-LM, September 2023
29. Frantar, E., Ashkboos, S., Hoefler, T., Alistarh, D.: GPTQ: accurate post-training quantization for generative pre-trained transformers (2023)
30. Pinecone Systems: Pinecone (2022). Accessed 01 Nov 2023
31. Reimers, N., Gurevych, I.: Sentence-BERT: sentence embeddings using Siamese BERT-networks. In: Proceedings of the 2019 Conference on Empirical Methods in Natural Language Processing. Association for Computational Linguistics, November 2019
32. Jiang, A.Q., et al.: Mixtral of experts (2024)
33. Shankar, S., Zamfirescu-Pereira, J.D., Hartmann, B., Parameswaran, A.G., Arawjo, I.: Who Validates the Validators? Aligning LLM-Assisted Evaluation of LLM Outputs with Human Preferences, April 2024

Enhancing Product Lifecycle Efficiency: Harnessing Natural Language Processing for Materials Insight and Optimization

Inés Pérez Couñago[✉], Lara Suárez Casabiell, Andrea Gregores-Coto, Christian Eike Precker, and Santiago Muiños-Landin

AIMEN Technology Centre, C/Relva, 27 A. Torneiros, 36410 O, Porriño, Pontevedra, Spain
ines.perez@aimen.es

Abstract. Materials play a pivotal role in manufacturing, serving as the foundation upon which the functionality and overall quality of products are built. In the material science domain, an overwhelming amount of knowledge is generated and stored as text encoding a humongous amount of information related to materials performance along the product life cycle that results fundamental in the manufacturing landscape, addressing adaptability and circularity . This study explores the application of Natural Language Processing techniques to analyze data availability, with a specific focus on the domain of polyvinyl chloride materials across the chemical, environmental, health, social and economic dimensions. While acknowledging the expanse of available academic data, this research also ventures into exploring vast web platforms, not often emphasized in the existing literature. Latent Dirichlet Allocation is employed to autonomously extract interconnected topics from textual data, providing a flexible tool to structure multifaceted datasets. Furthermore, its integration with question-and-answer schemes, powered by Large Language Models, represents a step forward in comprehensive data mapping. This combination aids in expediting the extraction of relevant information while contributing to the creation of a structured database where all relevant information pertaining to a particular topic is organized, identifying specific missing data or noncorrelated information. This approach promises to contribute to the evolution of data analysis methodologies, offering insights into the data landscapes of material science with impact in the current manufacturing scenario.

Keywords: materials · natural language processing · product lifecycle· sustainability

1 Introduction

In today's fast-evolving manufacturing sector, a diverse array of information is crucial for enhancing process efficiency [9]. Moreover, current regulations require the adoption of Life Cycle Assessment (LCA) strategies [14], which depend on extensive data regarding material value chains and their social and environmental impacts. The landscape of data sources is complex, encompassing databases, scientific literature and industry reports. Time and resource constraints, as well as data availability, significantly impact the manufacturing process.

© The Author(s) 2025
K. Alexopoulos et al. (Eds.): ESAIM 2024, LNME, pp. 234–241, 2025.
https://doi.org/10.1007/978-3-031-86489-6_24

Data-driven models have the potential to significantly accelerate the design of new materials [2]. Large Language Models (LLMs) facilitate the analysis of material properties [15], the retrieval of maintenance logs [10], and the creation of recommendation systems to enhance management [11]. However, AI tools that consider LCA in product design are not fully exploited [1]. Additionally, LLMs can produce inaccurate results in knowledge-intensive tasks due to their training on generic datasets [7]. Although the Retrieval-Augmented Generation (RAG) method addresses this issue [7], it still suffers from inaccuracies due to the presence of excess irrelevant data within the dataset.

This work aims to present a methodology to facilitate information retrieval within the manufacturing value chain, taking into account socioeconomic and safety impacts to provide valuable information for decision-making. The proposed approach integrates RAG with Topic Modeling for filtering out non- relevant data and improving computational efficiency. This article is structured as follows: The introduction section outlines the motivation behind the problem and presents the foundations of the retrieval tool. This is followed by the Methodology section, which details the technical approach, and the Results section. Finally, the paper concludes with the conclusions and future steps.

2   Methodology

We present a pipeline for creating a knowledge resource tailored to specific industries, illustrated through a case study in PVC. This pipeline delivers customized responses that address socioeconomic, chemical, and safety aspects. The structure of the tool is depicted in Fig. 1.

2.1   Data Acquisition

Data sources were identified from open-access bibliographic repositories and large online databases. Bibliographic information was extracted using the *UnstructuredFileLoader* module of the *Langchain* library, which is designed to handle PDF complex layouts. For web data extraction, *Selenium WebDriver* was used to simulate user actions and retrieve dynamically loaded content; the fetched URLs were then parsed with *BeautifulSoup* library. Tabular data was preprocessed by concatenating each table's header with its corresponding cells, creating a more comprehensive textual representation to help NLP techniques understand column correlations. Data is stored in a JSON file with the format data: extracted content, metadata: reference to ensure traceability. Ethical considerations were taken into account throughout this process, ensuring that no user-specific details were stored.

2.2   Topic Modelling

The Latent Dirichlet Allocation (LDA) model was used for topic modeling [12]. The implementation was carried out with the *Gensim* library, which required specifying the number of topics and control parameters affecting the specificity of topic distributions. We set these parameters to 0.1 to enhance topic distinctiveness and manually assessed the model's performance across 2, 3, and 5 topics. Only nouns, verbs, and adjectives

Fig. 1. Structure of the Information Retrieval Tool

were retained, and lemmatizing was applied to preserve semantically significant words. Words that appear in over 60% of documents were filtered out to exclude overly common terms. Additionally, terms related to web page characteristics, such as "cookies", were removed and varia tions of PVC (e.g., 'poly', 'vinyl', and 'chloride') were standardized to improve consistency in the analysis.

2.3 Retrieval

Retrieving information from a textual database involved several key NLP steps.

- **Splitter.** Textual data was divided into smaller fragments to fit within the model's context window. *RecursiveCharacterTextSplitter* from the *Langchain* library was used, setting a fragment length of 900 tokens (approximately 120 words) with 100 overlapping tokens.
- **Embedding.** The fragmented data and query were transformed into vector representations using the *XL-Instructor* embedding model, chosen for its open-source nature and strong performance in embedding benchmarks.
- **Vector store.** The *Chroma* vector store was employed to manage the storage of embedded data. We chose *Chroma* for its open-source nature and ability to operate entirely on a local machine.
- **Retriever.** Similarity between the query embedding and those in the vector store was computed using cosine similarity, noted for its simplicity and effectiveness. We retrieved the three most similar results, prioritizing longer, more relevant documents over a diverse set of shorter, less related ones.

2.4 Answer Generation

A generative LLM was deployed to generate responses prompted by the retrieved data. We chose the 7B version of the *Llama2* model, fine-tuned for chat and provided by Meta, due to its open-source nature, suitable window size, and compatibility with local GPUs.

2.5 Evaluation

The performance of the retrieval was assessed using various approaches: (i) statistical scorers that measure token matches between phrases (e.g., BLEU, METEOR, ROUGE) [3], (ii) NLP-based scorers that compare the semantic meaning of sentences (e.g., BERT Score, BLEURT) [13], and (iii) scorers utilizing an LLM to evaluate content adequacy according to a user-provided rubric (e.g., Prometheus) [6]. The six scorers were applied to compare the retrieved information with a manually defined reference. As it is not always possible to generate a reference answer, it was also compared the retrieved information directly with the question posed. Concordance between the scorers and human judgment was evaluated using Spearman's and Kendall's correlations [8].

3 Results

3.1 Data Extraction

Data on the PVC value chain was gathered from 67 sources, including blogs, Comptox and PubChem platforms, reports, and research papers. To explore the effect of noisy data, we included 10 sources related to the welding industry.

3.2 Topic Modeling

LDA implementation with 2 topics identified two distinct groups in the dataset: one related to the welding industry and the other to PVC. Increasing the number of topics to 3 retained the welding topic but split the PVC topic into two indistinct subtopics. With 5 topics, results became mixed. This demonstrates that LDA, when properly tuned, can effectively detect distinct themes and filter out irrelevant data.

Applying LDA exclusively to PVC data resulted in topics that combined chemical, economic, and safety terms. While this approach provided insights into dataset patterns, it was insufficient for hierarchical clustering. To achieve clearer topic definitions, we trained the LDA model on a curated subset of sources focused on chemical, safety and economic data and then applied it to the entire PVC dataset. This revised method yielded well-defined topics (see Fig. 2). To measure data availability for each dimension objectively, the dataset was divided into three groups, each containing sources where the corresponding topic was present in at least 60% of the data. The analysis showed that chemical data was predominant and socioeconomic data was the least represented, highlighting the challenge of accessing such information. Safety documents frequently appeared in conjunction with other topics.

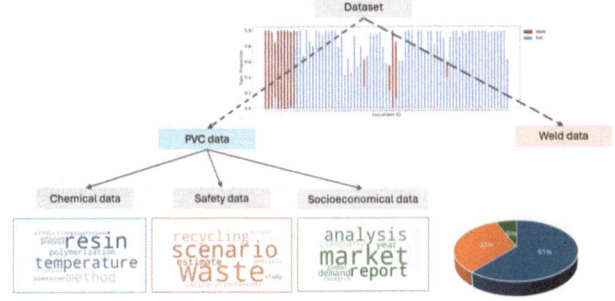

Fig. 2. Definition and distribution of topics in the dataset related to PVC.

3.3 Retrieval and Answer Generation

The information retrieval process was evaluated using the query "What is the melting point of PVC?" across different datasets: the entire dataset (PVC and welding data, composed of 77 data sources), the PVC-specific dataset (66 data sources), and subsets classified by the LDA model (chemical, safety, and economic). The results are summarized in Table 1. The pipeline successfully generated correct answers when the dataset included chemical data, indicating that the retrieval system effectively located documents related to thermal properties. In contrast, the economic and hazards subset contributed only noise, as it contained no relevant documents about PVC's thermal properties, resulting in a poor answer to the query.

Execution time was higher with the larger dataset. Removing the 10 noisy sources reduced the execution time by 6%. Further, filtering the dataset to include only documents with at least 60% relevance to the chemical topic reduced the execution time by 94.8%. The execution time for the safety-related dataset was unexpectedly high. This dataset was notably diverse, as its documents mixed content related to chemical properties and socioeconomic impacts. This suggests that diversity in the dataset, beyond just its size, significantly affects embedding and retrieval efficiency. Although this was a small-scale test, it indicates that larger and more diverse datasets will require more sophisticated retrieval strategies. The LDA model's filtering capabilities can enhance efficiency by reducing irrelevant data.

Table 1. Generated answers for the question "What is the melting point (m.p.) of PVC?" across different dataset sizes with execution times

Dataset	N° sources	Execution time (s)	LLM answer
PVC + weld	77	711	the m.p. is around 100–260 °C, but can vary depending on the specific formulation of the PVC
PVC	66	668	the m.p. is around 100–260 °C, but can vary depending on the specific formulation of the PVC
Chemical	25	36.6	the m.p. is around 100–260 °C, but can vary depending on the specific formulation of the PVC
Socioeconomic	15	5.07	the m.p. of PVC is likely to be influenced by the price of crude oil
Safety	2	470	PVC is a thermoplastic material that can be melted and reformed multiple times

3.4 Evaluation

The retrieved information was compared to a reference text using the scorers from the Methodology. Metrics rated the Chemical Topic highest for relevance on melting point data. Although subtle, the Hazards Topic, which included discussions of chemical properties, was more relevant than the Economic Topic. A direct comparison with the query is used as an alternative when reference answers are unavailable. This approach consistently showed that the Chemical Topic scored highest, while the Socioeconomic and Safety data were rated similarly.

Spearman's and Kendall's correlations revealed that NLP-based scorers were consistent with human judgment when compared to a reference response; however, their consistency may be reduced when lacking this ground truth. METEOR and Prometheus showed the most robust correlation with human judgment, even without a defined reference. Figure 3 illustrates the results.

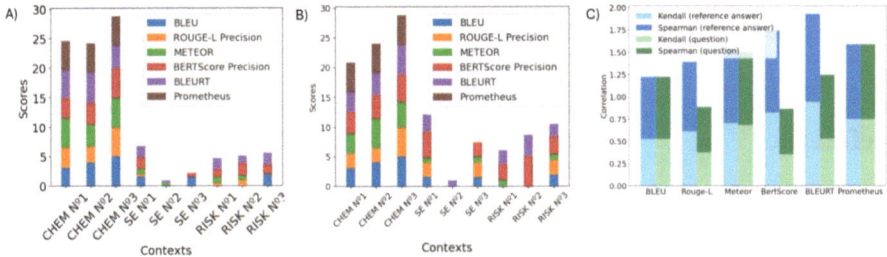

Fig. 3. Evaluation of the information retrieved for each topic (A) with the reference answer, (B) with the query and (C) Spearman's and Kendall's correlations.

4 Conclusions and Future Work

In this paper, we present the design of a tool to efficiently extract information from an extensive database. The proposed framework comprises three key components: (i) topic modeling to tailor the dataset according to the query topic, (ii) retrieval of information relevant to a user query, and (iii) generation of an answer based on the retrieved information. The pipeline was applied to retrieve information about the PVC value chain and LCA, providing users with easy access to relevant sources to expand on the provided information. Integrating LDA into the retrieval process significantly improved pipeline efficiency and reduced execution time, especially when handling diverse and broad datasets. This open-source solution ensures data privacy, offering an alternative to commercial options such as the OpenAI API.

This work also aims to emphasize the importance of developing new strategies to leverage the vast amount of information available in web content. This area remains relatively unexplored, with academic papers typically constituting the bulk of datasets used for retrieval tests. Academic sources often raise more significant ethical and legal concerns related to data privacy, security, and intellectual property. In contrast, responsibly using web data can be more straightforward to assess, as web content is generally more public and accessible, reducing the complexities associated with sensitive academic data.

Future research will focus on several key areas: automating parameter tuning for LDA, incorporating quantitative metrics to enhance reproducibility, and studying noise reduction techniques to improve dataset quality. We will also explore scalable NLP techniques, such as neural topic modeling, to handle more complex datasets efficiently. Evaluating the generation capabilities of LLMs will be prioritized, utilizing a broader range of evaluation metrics and user satisfaction surveys to provide a comprehensive assessment of the system's effectiveness in managing the inherent subjectivity of these tasks. Additionally, optimizing retrieval keys will be essential to maximizing the effectiveness of information extraction. While this study focused on the PVC value chain, the proposed framework will be adapted for other domains, providing valuable insights for sustainability assessments and decision-making in the manufacturing sector.

Aknowledgment. This work has been developed within the framework of the ANALYST project, which has received funding from the European Union's Horizon Europe Research and Innovation Program under grant agreement no. 101138548.

References

1. González-Val, C., Muiños-Landin, S.: Generative design for social manufacturing. Zenodo (2020). https://doi.org/10.5281/zenodo.4597558
2. Gregores-Coto, A., et al.: The use of generative models to speed up the discovery of materials. Comput. Meth. Mater. Sci. **23**(1), 14 (2023)
3. Hu, T., Zhou, X.H.: Unveiling LLM evaluation focused on metrics: challenges and solutions (2024). arXiv preprint arXiv:2404.09135. https://doi.org/10.48550/arXiv.2404.09135
4. Johri, P., Khatri, S.K., Al-Taani, A.T., Sabharwal, M., Suvanov, S., Kumar, A.: Natural language processing: History, evolution, application, and future work. In: Proceedings of the 3rd International Conference on Computing Informatics and Networks: ICCIN 2020, pp. 365–375. Springer Singapore (2021). https://doi.org/10.1007/978-981-15-9712-1_31
5. Jones, K.S.:. Natural language processing: a historical review. In: Current Issues in Computational Linguistics: in Honour of Don Walker, pp. 3–16. Routledge (1994)
6. Kim, S., Shin, J., Cho, Y., Jang, J., Longpre, S., Lee, H., Seo, M.: Prometheus: inducing evaluation capability in language models. In: NeurIPS 2023 Workshop on Instruction Tuning and Instruction Following (2023). https://arxiv.org/abs/2311.06941
7. Lewis, P., et al.: Retrieval-augmented generation for knowledge-intensive NLP tasks. Adv. Neural Inf. Process. Syst. **33**, 9459–9474 (2020). https://doi.org/10.48550/arXiv.2005.11401
8. Liu, Y., Iter, D., Xu, Y., Wang, S., Xu, R., Zhu, C.: GPTEval: NLG evalu- ation using GPT-4 with better human alignment (2023). arXiv preprint arXiv:2303.16634. https://doi.org/10.48550/arXiv.2303.16634
9. Liu, Y., Zhao, T., Ju, W., Shi, S.: Materials discovery and design using machine learning. J. Materiomics (2017). https://doi.org/10.1016/j.jmat.2017.08.002
10. May, M.C., Neidhöfer, J., Körner, T., Schäfer, L., Lanza, G.: Applying natural language processing in manufacturing. Procedia CIRP **115**, 184–189 (2022)
11. Melo, G., Chaves, M., Kolter, M., Schleifenbaum, J.H.: Skills requirements of additive manufacturing - a textual analysis of job postings using natural language processing. In: Klahn, C., Meboldt, M., Ferchow, J. (eds.) Industrializing Additive Manufacturing. AMPA 2023.

Springer Tracts in Additive Manufacturing, pp. 299–316. Springer, Cham (2023). https://doi.org/10.1007/978-3-031-42983-5_21

12. Neishabouri, A., Desmarais, M.C.: Reliability of perplexity to find number of latent topics. In: The Thirty-Third International Florida Artificial Intelligence Research Society Conference (2020)

13. Sellam, T., Das, D., Parikh, A.P.: BLEURT: Learning robust metrics for text generation (2020). arXiv preprint arXiv:2004.04696. https://doi.org/10.48550/arXiv.2004.04696

14. Stavropoulos, P., Giannoulis, C., Papacharalampopoulos, A., Foteinopoulos, P., Chryssolouris, G.: Life cycle analysis: comparison between different methods and optimization challenges. Procedia CIRP **41**, 626–631 (2016). https://doi.org/10.1016/j.procir.2016.01.014

15. Swain, M.C., Cole, J.M.: ChemDataExtractor: a toolkit for automated extraction of chemical information from the scientific literature. J. Chem. Inf. Model. **56**(10), 1894–1904 (2016)

Prompting to Gather Object Categories in NeRF Scenes Related to Manufacturing

Selen Pehlivan[(✉)] and Santeri Hyvärinen

VTT Technical Research Center of Finland, Oulu, Finland
selen.pehlivantort@vtt.fi

Abstract. Despite the effectiveness of closed-set object detectors, recent advancements have introduced zero-shot detectors that can recognize a wide range of object categories across different environments. These detectors rely on text prompts, such as object tags. This study explores using multimodal large language models (MLLMs) to gather and refine object information from NeRF scenes into tags. We propose a training-free pipeline for extracting object-specific details, such as category, color, material, and functionality, from 3D scenes via prompting. Subsequently, we investigate how to apply the object tagging problem to NeRF-reconstructed scenes, particularly in a manufacturing context. This pipeline is evaluated in manufacturing environments for object recognition, with the resulting categories serving as inputs for zero-shot object detection and other tasks.

Keywords: Multimodal Large Language Models · 3D Scene Understanding · NeRF · Object Recognition · Prompting

1 Introduction

In industrial settings such as factories, 3D scene understanding enables intelligent agents to execute complex tasks. For instance, factory robots can recognize and interact with objects, navigate the factory floor, and assist in assembly tasks. This technology holds significant promise for various applications, including object inspection, safety enhancement, layout design, and production process optimization [1,2]. Our study focuses on exploring multi-label object recognition within the context of 3D scene understanding.

Multi-label object recognition for object tagging is essential in computer vision for tasks like object detection and segmentation. In particular, open-vocabulary pipelines, such as zero-shot object detectors, support a much wider label set than closed-set ones, and a tagging system can assist users in those pipelines by offering natural text prompts as labels. Advancements in vision-language models, e.g., CLIP [3], and their integration with large-language models (LLMs) [4] have enabled to recognize common objects in images with zero-shot capabilities, as shown in recent studies [5,6]. However, the recognition problem

K. Alexopoulos et al. (Eds.): ESAIM 2024, LNME, pp. 242–250, 2025.
https://doi.org/10.1007/978-3-031-86489-6_25

has primarily been studied for interpreting objects in images. Additionally, Multimodal Large Language Models (MLLMs) like LLaVA [7] and GPT-4V [8], which demonstrate strong reasoning abilities with both language and image prompts, can be used for tagging without training.

Following this, our study explores the use of LLaVA-NeXT [9] for object recognition in 3D scenes. While most research focuses on image-based models, 3D visual modalities are less studied. We examine Multimodal Large Language Models (MLLMs) for recognizing objects in 3D scenes, leveraging their language capabilities to categorize objects with attributes across different views using a training-free approach. By various prompts (see Fig. 1), we gather rich information including various attributes from multiple scene perspectives.

As one of our target, the tagging solution is integrated into a web-based application to investigate the environment, particularly the clutter one. Modules that refine tag proposals according to users' search criteria, such as objects by functionality or material or categorizing items as damaged are also integrated into the interface. This interface can be embedded in intelligent systems as a monitoring tool to improve automated manufacturing processes with increased user interaction or as an annotation tool for collecting image-text data pairs. It is worth noting that our study presents preliminary findings from a pipeline still under development.

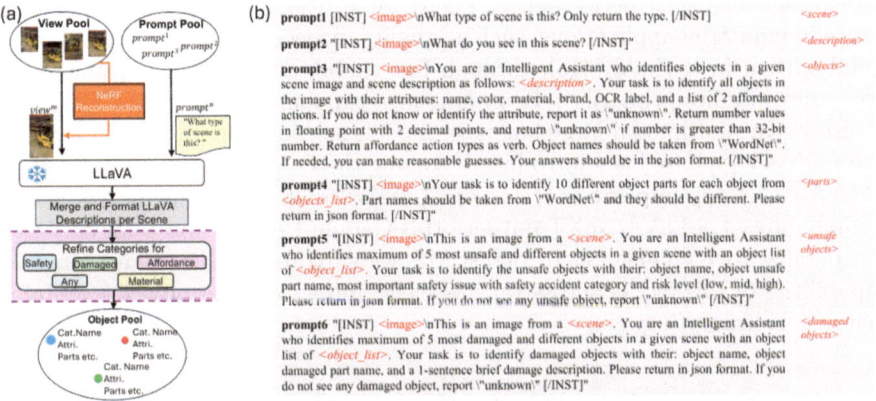

Fig. 1. Collecting objects through MLLMs from views of 3D scenes: (a) shows our pipeline for querying a MLLM and (b) shows our prompts.

2   Background

This section first gives the basics of NeRF representation and then briefly presents an overview of large multimodal models.

Neural Radiance Field (NeRF). NeRF [10] is a technique for novel view synthesis, encoding scenes as continuous volumetric radiance fields. These fields return density σ and color c for each 3D point p and viewing direction d, allowing rendering of 2D images for rays r with color values of

$$I'(r) = \int_{t_n}^{t_f} T(t)\sigma(t)c(t)dt, \qquad (1)$$

where $T(t) = exp\left(-\int_{t_n}^{t} \sigma(s)ds\right)$ with near and far bounds t_n and t_f. During training, the rendered values, $I'(r)$, is compared to the ground truth pixel values, $I(r)$, with a rendering loss $L = \sum_{r \in \mathcal{R}} ||I'(r) - I(r)||_2^2$ where $\mathcal{R}$ is the set of rays.

NeRF is less commonly used in industrial applications compared to point cloud representations, with only a few basic attempts [11]. In selecting NeRF for scene modeling, we aim to empirically test its effectiveness when integrated with MLLM for our task.

Large Multimodal Models. Our work leverages the latest advancements in Large Language Models (LLMs) such as GPT [4] and LLaMA [12], known for their exceptional generalization abilities and use as universal interfaces for various language tasks with a single model. Instruction-tuned versions like Chat-GPT [4] and InstructGPT [13] are developed through fine-tuning or involving humans in the loop. While LLMs have strong reasoning abilities, they may struggle in task-specific scenarios. Fine-tuning and chain-of-thought (CoT) prompting [14] enhance their performance. There are also some recent studies utilizing LLMs in industrial applications, such as quality inspection [15] and human-robot collaboration in assembly [16], through fine-tuning and CoT prompting. However, these studies primarily rely on language capabilities.

Recently, multimodal large language models such as Gemini [17], LLaVA [7], and GPT-4V [8] have been introduced. Among these, we employ the open-source LLaVA-NeXT [9], improved version of LLaVA-1.5 [7]. Despite being trained on a small dataset, LLaVA-NeXT outperforms Gemini Pro [17] and achieves scores comparable to GPT-4V [8]. In our study, we explore the vision and language capabilities of this model across various scene views to gather object-level information via prompting.

3 Our Approach for Collecting Objects

There exist strong image-based 2D foundational models that rely on billion-scale datasets. To leverage these models for 3D scene interpretation, Neural Radiance Field (NeRF), which connects multiple 2D views into a 3D scene via a deep neural network with 2D rendering capability, is used as the 3D representational model in this study.

Both the captured images to reconstruct NeRF and the rendered images from NeRF can be directly input into MLLMs like LLaVA, since these models accept 2D images. We first render multiple images from the NeRF model of a scene and then collect categories and object-level information from LLaVA via prompting.

3.1 Prompts for Multimodal Large Language Model

A set of prompts is defined to gather high-level object-related information from multiple rendered images of NeRF scenes using MLLMs. By querying these prompts to LLaVA-NeXT, a state-of-the-art visually-assisted large language model, we request the following information for each scene view: the category of the $< scene >$, a $< description >$ of the scene, $< objects >$ with attributes and parts, $< unsafe\ objects >$, and $< damaged\ objects >$. Given in Fig. 1, the prompt summaries are as follows:

- Given an $< image >$, **prompt1** and **prompt2** asks MLLM to provide with the $< scene >$ category and a $< description >$ of the scene, respectively.
- Given the $< image >$ and $< description >$, **prompt3** asks MLLM to generate a list of scene $< objects >$ using *WordNet*. This list includes each object's name, color, material, brand, OCR label, and two affordance actions.
- Given the $< objects >$, **prompt4** asks MLLM to return 10 object parts for each object category from the object list using *WordNet*.
- Given the $< image >$, $< scene >$ category, and $< objects >$, **prompt5** asks MLLM to identify the five most unsafe objects, along with their names, the unsafe parts, and the two most important safety issues.
- Given the $< scene >$ category and $< objects >$, **prompt6** asks MLLM to identify the five most damaged objects, providing the name of each object, the damaged part, and a brief description of the damage.

Leveraging the capabilities of a multimodal model with both visual and textual inputs, prompts gather object knowledge across views of a 3D scene. Rather than fine-tuning the model, the pre-trained model weights are used. To ensure consistency in object category names, object lists are requested from *WordNet* [18] (see **prompt3-4**). Due to MLLM processing time, prompting is applied in an offline setting where all images for NeRF construction are queried with a set of predefined prompts in the initial experiments.

3.2 Aggregated Object Categories Across Views

By collecting sets of objects with attributes for each view of the 3D scene, the object lists from the images are then combined into a unified object category list. Our scene tags specifically encompass all object categories. We conduct experiments on both the images captured for NeRF reconstruction and the sampled views rendered by the NeRF model. These aggregated categories serve as input for downstream tasks, such as open-vocabulary object detection, to perform per-image object detection on the rendered images.

4 Experiments

Datasets. Two small-scale datasets, the Replica [19] including high-quality indoor scene scans with ground-truths, and our data collection in the manufacturing context, are used during our experiments. Scenes from our collection

Fig. 2. Our manufacturing-related scenes via (a) Vanilla-NeRF and (b) DvGO.

cover uncommon object categories such as various robots and assembly tools (see Fig. 2). First, we use pre-rendered Replica data of 8 scenes with 2D instance labels provided by Zhi et al. [20] to evaluate recognition performance. Later, some qualitative experiments are conducted on our data collection.

NeRF Results. We employed various NeRF reconstruction techniques, including vanilla NeRF [10] and the accelerated Direct Voxel Grid Optimization (DvGO) [21]. Figure 2 presents reconstructions of our manufacturing-related captures and includes rendered images from challenging viewpoints via DvGO.

Recognition Results. Category names for querying are often limited to a pre-defined vocabulary, as traditional systems trained on closed-label sets may not support categories beyond this range. Additionally, users may have difficulty identifying category names in natural text, even if the system supports an open vocabulary. Given the diverse range of object categories in the scenes, a comprehensive list of image tags is advantageous for enabling effective user searches.

Our LLaVA-based solution is compared with the Recognize Anything Plus Model (RAM++) with swin-large 14M version [6] on both pre-rendered [20] and NeRF-rendered Replica scenes. RAM++ is a recent image tagging system that accurately identifies a wide range of categories using large-scale image-text pairs for training rather than relying on manual annotations. While RAM++ is image-based, it is the most relevant comparison for our study, as we propose a MLLM-based tagging approach for NeRF-generated images.

Although semantic labels are available as ground-truth for the Replica dataset, there are discrepancies between the predicted object categories from LLaVA (or RAM++) and the Replica ground-truth categories. The object tags proposed by the models are quite diverse, offering a wide range of suggestions, including synonyms and multi-word phrases. To evaluate recognition ·performance, we employed a strategy using LLMs. We first collected tags from various scene views using RAM++ and our method. These tags were then aggregated into tag sets per scene for each method. Next, we used LLaVA to map LLaVA-tags and RAM-tags, respectively, as the prediction list $< prelist >$ to Replica-tags as the ground-truth list $< gtlist >$ using the following prompt: *For* *a category name in the prediction list, find the synonymous category from the ground-truth list. Make the most reasonable guess. If there is no match, map to "unknown". The predicted categories list is < prelist >, and the ground-truth category list is < gtlist >. Please return in a json format where the keys are ground-truth categories.*

Tag prediction results using our model with the LLaVA-7B, LLaVA-13B, and RAM++ are detailed in Table 1. There is not a standard metric to apply for evaluation. Our evaluation metric for recognition performance involves the number of shared tags that are calculated between the predicted-tags (LLaVA-tags or RAM-tags) and ground-truth tags (Replica-tags). This count is then normalized by the total number of Replica-tags per scene and scaled over 100%. This is the percentage of relevant predicted tags over ground-truth tags. Despite promising performance of LLaVA, we noted a significant decline in tag generation for NeRF-rendered images due to image quality issues. The performance of all models on pre-rendered images consistently exceeds that on NeRF-rendered images. This shows the prediction accuracy is strongly affected by the reduced image quality in some viewpoints after the reconstruction. Moreover, we observed cases of failed NeRF reconstruction, such as *office3*.

Table 1. Object category recognition performance using (a) RAM++, (b) ours with LLaVA-7B and (c) ours with LLaVA-13B.

Pre-rendered images [20]								
Model	office 0	office 1	office 2	office 3	office 4	room 0	room 1	room 2
ram++	37.04	45.83	50.00	65.52	68.42	50.00	62.5	56.52
llava7b	59.26	54.17	66.67	60.00	68.42	100.0	87.5	77.27
llava13b	70.37	50.00	70.83	70.00	78.95	74.07	87.50	77.27

NeRF-rendered images								
Model	office 0	office 1	office 2	office 3	office 4	room 0	room 1	room 2
ram++	51.85	00.00	29.17	–	31.58	100.0	37.50	40.91
llava7b	29.63	41.67	37.50	–	57.89	60.71	75.00	39.02
llava13b	37.04	45.83	20.97	–	47.37	100.0	58.33	63.64

Qualitative Results. Along with recognition evaluation, we conducted qualitative experiments on our data collection, including manufacturing samples. We developed an application that integrates our LLaVA-based tagging solution with downstream tasks and the NeRF reconstruction pipeline (see Fig. 3(a)). This tool includes an interactive interface that allows users to filter object tags by criteria like affordance (e.g., *sitting*) and select subsets for zero-shot tasks. Grounding-DINO [22] localizes objects by creating bounding boxes using provided tags, while Segment Anything [23] returns object masks.

Our first qualitative evaluation focused on generating object summaries, as illustrated in Fig. 3(b). After collecting object details from multiple images, we used LLaVA to generate summaries for the selected category. Our second evaluation tested downstream tasks, including open-vocabulary object detection with the recent Grounding-DINO [22], as given in Fig. 3(c). By providing a list of object tags and an image, this detector localizes objects identified by our tagging solution. Please note that the user interface supports the selection of a small tag subset from this list.

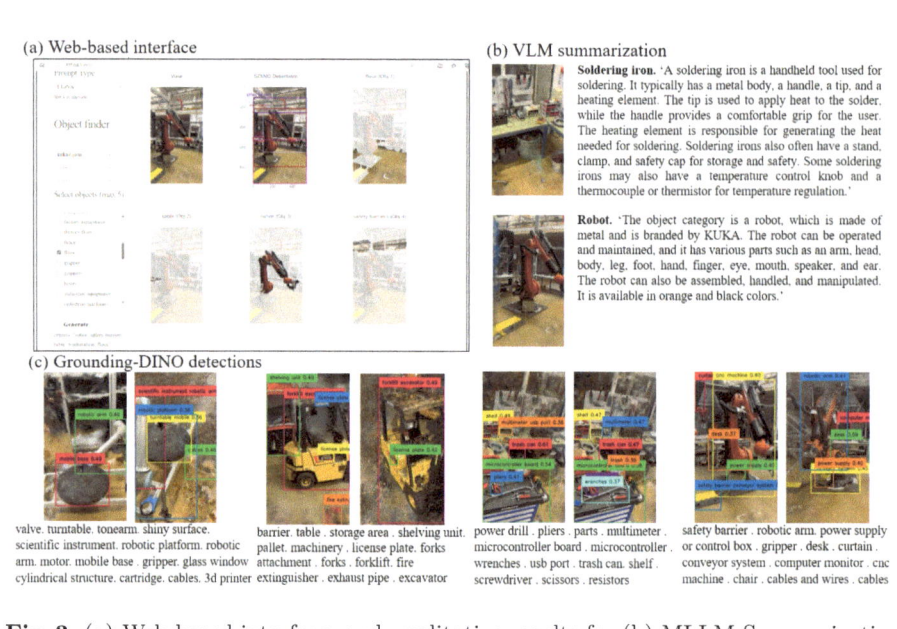

Fig. 3. (a) Web-based interface, and qualitative results for (b) MLLM Summarization and (c) Grounding-DINO detections.

5   Conclusion

Our study introduces prompts for querying multimodal large language models and presents preliminary tag results for NeRF scenes by leveraging their visual and language reasoning capabilities. LLaVA-NeXT has been used to generate object tags with semantic information. The proposed solution aims to reduce the need for human labeling when integrated into an annotation tool, or enable zero-shot capabilities for open-vocabulary tasks. Our pipeline has been integrated into a web-based application that can be used as (1) an annotation tool to obtain text-image pairs and (2) a monitoring tool attached to intelligent agents.

Our preliminary findings suggest that the object tag proposals are in natural language form, providing a range of language expressions. Although the current tag list covers a broad range of object categories, it is extensive and could be improved through further refinement using a fine-tuned model in future work.

Acknowledgement. This research funded by the VTT Technical Research Centre of Finland. We thank the CSC-IT Center for Science for computer resources.

References

1. Fan, J., et al.: Vision-based holistic scene understanding towards proactive human-robot collaboration. Robot. Comput.-Integr. Manuf. **75**, 102304 (2022)
2. Riaz, H., Terra, A., et al.: Scene understanding for safety analysis in human-robot collaborative operations. In: International Conference on Control, Automation and Robotics (2020)
3. Radford, A., Kim, J.W., et al.: Learning transferable visual models from natural language supervision. In: ICML (2021)
4. Achiam, J., Adler, S., Agarwal, S., et al.: GPT-4 technical report. arXiv preprint arXiv:2303.08774 (2023)
5. Zhang, Y., Huang, X., Ma, J., et al.: Recognize anything: a strong image tagging model. arXiv preprint arXiv:2306.03514 (2023)
6. Huang, X., Huang, Y.-J., et al.: Open-set image tagging with multi-grained text supervision. *arXiv e-prints*, pages arXiv–2310 (2023)
7. Liu, H., Li, C., Wu, Q., Lee, Y.J.: Visual instruction tuning. In: NIPS, vol. 36 (2024)
8. Wu, Y., Wang, S., Yang, H., et al.: An early evaluation of GPT-4v(ision). arXiv preprint arXiv:2310.16534 (2023)
9. Liu, H., Li, C., Li, Y., Li, B., Zhang, Y., et al.: LLaVA-next: improved reasoning, OCR, and world knowledge (2024)
10. Mildenhall, B., et al.: NeRF: representing scenes as neural radiance fields for view synthesis. Commun. ACM **65**(1), 99–106 (2021)
11. Šlapak, E., Pardo, E., Dopiriak, M., et al.: Neural radiance fields in the industrial and robotics domain: applications, research opportunities and use cases. arXiv preprint arXiv:2308.07118 (2023)
12. Touvron, H., Lavril, T., Izacard, G., et al.: LLaMA: open and efficient foundation language models. arXiv preprint arXiv:2302.13971 (2023)
13. Ouyang, L., Wu, J., Jiang, X., et al.: Training language models to follow instructions with human feedback. In: NIPS (2022)
14. Wei, J., Wang, X., Schuurmans, D., et al.: Chain-of-thought prompting elicits reasoning in large language models. In: NIPS (2022)
15. Zhou, B., Li, X., et al.: CausalKGPT: industrial structure causal knowledge-enhanced large language model for cause analysis of quality problems in aerospace product manufacturing. Adv. Eng. Inform. **59**, 102333 (2024)
16. Gkournelos, C., et al.: An LLM-based approach for enabling seamless human-robot collaboration in assembly. CIRP Ann. **73**, 9–12 (2024)
17. Gemini Team, Anil, R., et al.: Gemini: a family of highly capable multimodal models. arXiv preprint arXiv:2312.11805 (2023)
18. Miller, G.A.: WordNet: a lexical database for English. Commun. ACM **38**(11), 39–41 (1995)

19. Straub, J., Whelan, T., et al.: The replica dataset: a digital replica of indoor spaces. arXiv preprint arXiv:1906.05797 (2019)
20. Zhi, S., Laidlow, T., et al.: In-place scene labelling and understanding with implicit scene representation. In: ICCV (2021)
21. Sun, C., et al.: Direct voxel grid optimization: super-fast convergence for radiance fields reconstruction. In: CVPR, pp. 5459–5469 (2022)
22. Liu, S., et al.: Grounding DINO: marrying DINO with grounded pre-training for open-set object detection. arXiv preprint arXiv:2303.05499 (2023)
23. Kirillov, A., et al.: Segment anything. In: ICCV, pp. 4015–4026 (2023)

Fundamental AI Topics

Efficient MLOps: Meta-learning Meets Frugal AI

Eduardo Peixoto[2], Diogo Torres[1], Davide Carneiro[1,2(✉)], Bruno Silva[3], and Paulo Novais[4]

[1] INESC TEC, R. Dr. Roberto Frias, 4200 Porto, Portugal
davide.r.carneiro@inesctec.pt
[2] Escola Superior de Tecnologia e Gestão, Instituto Politécnico do Porto, Felgueiras, Portugal
[3] Muvu Technologies, Lisbon, Portugal
[4] ALGORITMI Research Centre/LASI, University of Minho, Braga, Portugal

Abstract. The advent of large Machine Learning models and the steep increase in the demand for AI solutions occurs at the same point in time in which policies are being enacted to implement more sustainable processes in virtually every sector. This means there is a need for more, better and larger models, which require significant computational resources, while at the same time a call for a decrease in the energy spent in the processes associated to MLOps. In this paper we propose a reduced set of meta-features that can be used to characterize sets of data and their relationship with model performance. We start from a large set of 66 features, and reduce it to only 10 while maintaining the strength of this relationship. This ensures a process of meta-feature extraction and prediction of model performance that is in line with the desiderata of Frugal AI, allowing to develop more efficient ML processes.

Keywords: Frugal AI · MLOps · Meta-Learning · Manufacturing

1 Introduction

In the last years, Machine Learning (ML) moved from being mostly a research topic, to being an applied and engineering one. The methodologies involved also changed accordingly, giving birth to what is nowadays known as Machine Learning Operations (MLOps) [5]. The field of MLOps is rapidly evolving to address the challenges of deploying and managing ML models in real-world scenarios. A particularly significant challenge arises when dealing with data streams that exhibit concept drift, in which the underlying relationship between the input features and the target variable changes over time [3]. This phenomenon can significantly degrade the performance of deployed models.

Traditional MLOps practices often rely on retraining models with the entire accumulated data, leading to increased computational costs, downtime during

© The Author(s) 2025
K. Alexopoulos et al. (Eds.): ESAIM 2024, LNME, pp. 253–261, 2025.
https://doi.org/10.1007/978-3-031-86489-6_26

redeployment, and potential storage bottlenecks. These issues become particularly acute for resource-constrained environments or applications processing high-velocity data streams.

In this context, the concept of Frugal AI has emerged as a promising paradigm for designing and deploying ML models in resource-scarce settings. Frugal AI [9] strives towards ML processes that are efficient in terms of computational power, memory usage, and data footprint. While this can be achieved in different ways (e.g. data minimization, transfer learning, active learning), this paper focus on the specific case of data streaming with concept drift. Specifically, we propose an approach to reduce the need for training new ML models by reusing previously trained models. To this end, we investigate a minimum set of meta-features to characterize the similarity of data blocks, as a proxy to select potentially similar models. Results show that reused models have, in average, a difference of 6% in accuracy, but require, in average, 17 times less time to be placed in production.

2   Contextualization

Frugal AI is a set of methodologies focused on developing resource-efficient and cost-effective AI solutions. This concept aligns with the principles of Green AI, which seeks to mitigate environmental impacts by optimizing algorithms, improving hardware efficiency, and adopting sustainable data management practices [2]. In the current context, where the demand for AI solutions is growing, coupled with increasing concerns over sustainability and efficiency, Frugal AI emerges as a crucial approach for balancing these divergent needs.

The computational side includes post-training optimization methods such as pruning (removing unnecessary parts of a model [6]), quantization (reducing the precision of parameters [6]), distillation (transferring knowledge from larger to smaller models), sparse representations, and low-rank factorization (decomposing matrices to reduce computation). It also involves leveraging pre-trained models and transfer learning to adapt models trained on large datasets to new tasks. Continuous learning techniques play a significant role as well, including active learning (prioritizing the most informative data points for training) and rotation-based methods like Local Fisher Discriminant Analysis (reducing dataset dimensions and improving model performance). Independent Component Analysis is another technique used to separate a multivariate signal into additive, independent components, further enhancing learning efficiency.

From the data perspective, Frugal AI employs data reduction techniques, including selection methods like stratified random sampling, replacement strategies, and transformation techniques such as PCA [4,7]. It also involves data generation approaches, which allow to minimize the resources and time spent in collecting and labeling data. These include data augmentation, synthetic/simulated data generation, and oversampling methods like SMOTE and ADASYN.

In this work, we achieve a Frugal AI solution through meta-learning. Meta-learning aims to improve the performance of ML algorithms by leveraging knowledge from previous ML tasks. This knowledge, encoded as meta-features, contains meta-characteristics of datasets and the resulting model performance metrics. These meta-features include simple statistical properties, advanced statistical measures, information-theoretic metrics, or model-based features [8].

3 Methodology

The methodology followed to find the reduced set of meta-features had two main phases (Fig. 1). It's main goal is to find a reduced set of meta-features that can be used for past model retrieval and reuse, thus minimizing the resources spent in training new models, in line with the Frugal AI view. In the first phase we selected 6 streaming datasets with concept drift, 3 of which synthetic. Each dataset was divided into a number of fixed-sized blocks, depending on the size of the dataset. For each block we extracted an initial set of 66 meta-features using the PyMFE library [1] and we also trained a Random Forest model using the default configuration. We extracted the main performance metrics of each model (e.g. Accuracy, precision, recall) and built 6 meta-datasets. Each of these meta-datasets has one row for each block of data in the original dataset, and makes the correspondence between the characteristics of that block of data (meta-features) and the quality of the model trained.

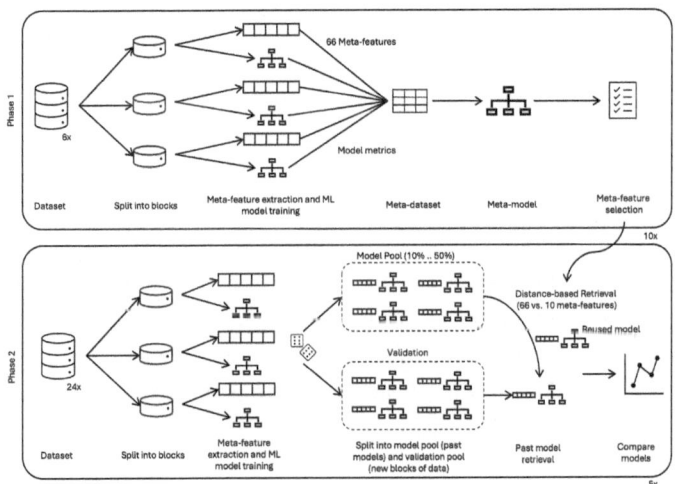

Fig. 1. Graphical representation of the methodology.

We then trained one Random Forest with each of the meta-datasets to predict accuracy (dependent variable) based on the meta-features (independent variables). These meta-models were trained with the purpose of calculating the relative importance of each meta-feature in predicting the accuracy of the corresponding model. After training each meta-model, we selected all the meta-features that had a relative importance of at least 1% (compared with the most relevant meta-feature). After doing this for the six meta-models, we selected the meta-features that were pre-selected as relevant in at least 3 out of the 6 cases.

Finally, and since Random Forests are ensembles of Decision Trees trained on subsets created by randomly sampling from the original dataset, we repeated this process 10 times, to rule out the effect of luck. This resulted in 10 reduced sets of meta-features that ranged between 5 (9th iteration) and 10 (4th and 6th iterations). The final set of 10 meta-features was selected according to their highest frequency over all 10 iterations. These are described in Table 1, in which the frequency quantifies the number of iterations in which this meta-feature was selected (out of 10).

In the second phase, the main goal was to validate the selected set of meta-features as appropriate to reuse models based on the distance between the underlying data. To this end 24 new datasets were used. Similarly to the first phase, each dataset was split into fixed-sized blocks, meta-features were extracted for each block (using both groups of 66 and 10), and a model was trained for each block. Blocks and models were then randomly split into two groups: pool and validation.

Table 1. Reduced set of 10 selected meta-features and their frequency over the 10 iterations of the process.

Meta-feature	Description	Frequency
nre	Normalized relative entropy	10
one_nn.mean	Performance of the 1-Nearest Neighbor classifier	9
freq_class.sd	Relative frequency of each distinct class	9
linear_discr.sd	Performance of the Linear Discriminant classifier	9
elite_nn.mean	Performance of Elite Nearest Neighbor	7
naive_bayes.sd	Performance of the Naive Bayes classifier	5
linear_discr.mean	Performance of the Linear Discriminant classifier	4
best_node.mean	Performance of the best single decision tree node	4
naive_bayes.mean	Performance of the Naive Bayes classifier	2
mean.sd	Mean value of each attribute (standard deviation)	2

For each block in the validation group (which represents an oncoming new block of data), we retrieved the closest block of data in the pool and the corresponding model. Distance was measured using the Bray-Curtis distance and considering the meta-feature vectors of both blocks, for 66 and 10 meta-features. This allows to compare how both groups of meta-features perform when used to select similar data. We then test both retrieved models (using 10 and 66 meta-feature) on the validation block of data, and compare their performance with that of the corresponding validation model. The goal is to retrieve (reuse) models that have a similar performance on the new data as a model that was specifically trained on that new data, and that is what is simulated and evaluated by this process. At the end we analyze the correlation between the performance of the validation models and the reused models, as well as the distance between the accuracy of each pair of models.

Finally, and since the performance of this approach also depends on the number of models on the pool and on their diversity, and since models are randomly selected to be in the pool, we ran these experiments with 5 different pool sizes (10% to 50%), and repeat each experiment 5 times to rule out the effect of luck. In the following section we analyze the performance of both sets of features in reusing models, and show that the smaller set of only 10 features has, in average, a similar performance, while being less computationally intensive.

For this research work to identify the most relevant meta-features of the data, the computational effort to find the similar model is smaller. This reduced set of meta-features can then be utilized to build more efficient and lightweight AI systems, aligning with the principles of frugal AI.

4   Results

This section presents the results of validating the methodology described above with a group of 24 datasets, which were not used when selecting the reduced set of meta-features. We do so based on two indicators: 1) the correlation between the accuracy of each model in the validation group and that of the model reused from the pool; and 2) the Mean Absolute Difference (MAD) between the accuracy of the models in the validation group and that of the reused models. The former quantifies how well aligned the retrieved models are with the actual models, whereas the latter reveals how distant they are in terms of performance. Table 2 shows the results of this analysis for the 24 datasets considered and a pool size of 20%.

In general, for the 20% pool size, using the selected 10 meta-features maintained a strong correlation compared to sets of 66 meta-features across a variety of datasets, in many cases even surpassing the full set (Table 2). This happens, for example, in the "Social_Network_Ads", "bank_10000" or "hyper_f" datasets. There are however a minority of cases in which using only 10 meta-features to reuse models resulted in worse correlation between models and/or increased error. This includes datasets such as "cardio_vascular_10000" and "pima_indians_diabetes" and points out that a case by case selection of meta-features should be done for each use case, following the proposed methodology.

When considering a pool size of 20%, the reduction to 10 meta-features outperformed sets of 66 meta-features concerning the correlation in 62.5% of the datasets, while obtaining a smaller MAD in 37.5% of the datasets.

Table 3 presents a comparative analysis of the performance of two sets of meta-features (66 and 10) at different pool sizes: 10%, 20%, 30%, 40% and 50%. The table shows the average values of correlation and MAD for both groups of meta-features, and a count of the percentage of datasets in which the models retrieved using 10 meta-features outperform those retrieved using 66. Looking at the average values, the 10 meta-features outperform the 66 in 2 out of the 5

Table 2. Performance Comparison between Meta-Features sets of 66 and 10.

Dataset	66 Meta-Features		10 Meta-Features	
	Correlation	MAD	Correlation	MAD
airlines	**0.85**	**0.04**	0.8	**0.04**
elec	0.54	0.06	**0.6**	**0.01**
general_data	0.32	0.07	**0.46**	**0.05**
german_credit_data	0.69	**0.05**	**0.8**	0.06
Social_Network_Ads	0.81	0.05	**0.91**	**0.01**
agr_a	0.84	**0.01**	**0.87**	0.01
cardio_vascular_10000	**0.74**	**0.01**	0.4	0.07
pima_indians_diabetes	**0.74**	**0.06**	0.48	0.06
sea_a	0.43	**0.01**	**0.85**	0.05
wpbc	**0.71**	0.13	0.07	**0.03**
abalone_dataset	0.42	**0.04**	**0.51**	0.19
cardiovascular_diseases_dv3	0.72	**0.01**	**0.73**	0.02
bank_10000	0.08	0.78	**0.85**	**0.01**
covtype	**0.57**	0.27	−0.12	**0.08**
winequality_red	**0.74**	**0.06**	0.34	0.11
creditcard_1000	0.37	0.03	**0.69**	**0.01**
winequality_green_4500	−0.31	0.08	**0.02**	**0.02**
Housing	−0.05	0.18	**0.79**	**0.01**
framingham_heart_disease	**0.71**	**0.02**	0.58	0.06
hyper_f	0.35	**0.02**	**0.74**	0.02
HR_Employee_Attrition	**0.93**	**0.01**	0.86	0.02
wdbc	0.2	**0.05**	**0.29**	0.38
world_food_scrubbed	−0.07	**0.03**	0.03	**0.09**
calories_9000	**0.73**	**0.01**	0.02	0.06

pool sizes considered, while this is true for 4 out of 5 when considering the MAD. When considering each dataset individually, the 10 meta-features outperform the 66 for the pool sizes of 10% and 20%, and match it for 50% (respectively 54%, 63% and 50%). In what concerns the MAD, the proposed set of meta-features outperforms the 66 in 38% to 46% of the cases.

We also analyzed the computational cost of the proposed approach. Specifically, we measured the average extraction time for the sets of 10 and 66 meta-features (Fig. 2(a)). It can be noted that a 6-fold reduction in the number of meta-features does not lead to a 6-fold reduction in the extraction time. This is explained by the nature of the 10 selected meta-features, which are mostly model-based, and computationally more expensive than other simpler meta-features, such as those based on statistics. However, the extraction time, whether for 10 (below 3 s in average) or 66 meta-features (below 5 s in average), is still much lower than that of training models, which takes 54.47 s in average. This means it takes, in average, 10.5 times longer to train the model than to extract the 66 meta-features, and 17 times longer than to extract the selected 10. This is thus a computationally less expensive operation that always training new models when new data is available.

Table 3. Summary of the results for different pool sizes when comparing 66 vs. 10 meta-features.

Pool Size	66 Meta-Features		10 Meta-Features		Outperforms	
	Corr.	MAD	Corr.	MAD	Corr.	MAD
10%	**0.45**	**0.067**	0.37	0.07	**54%**	38%
20%	0.50	0.09	**0.52**	**0.06**	**63%**	38%
30%	**0.53**	0.064	0.42	**0.062**	46%	33%
40%	0.45	0.062	**0.47**	**0.06**	46%	46%
50%	**0.58**	0.09	0.51	**0.06**	**50%**	46%

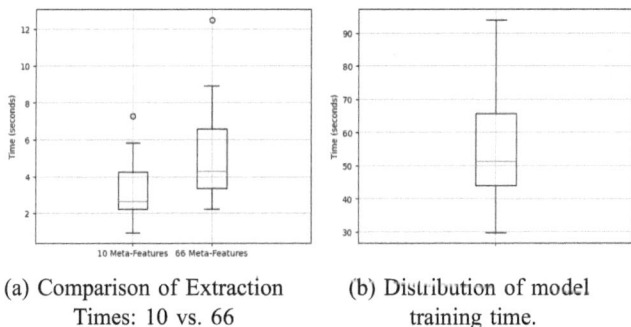

(a) Comparison of Extraction Times: 10 vs. 66 Meta-Features.

(b) Distribution of model training time.

Fig. 2. Distribution of the duration of extracting meta-features (10 vs. 66) and of training models.

5   Conclusions

In this article, we described an approach aiming to reduce the computational overhead of training new models in scenarios of streaming data with concept drift, based on the reuse of previously trained models, selected according to a reduced set of meta-features.

One key contribution of this paper is thus the reduced set of meta-features that can be used as a proxy for model performance. The fact that these meta-features were computed using a publicly available library makes these results reproducible and applicable by other researchers.

We also show that the computational cost of extracting these meta-features and reusing past models is, in average, 17 times smaller than that of training the actual models. This approach would thus scale better than traditional approaches, in which models are simply re-trained with the new data. These results were obtained from applying the proposed methodology to a set of 24 ML problems, which gives some confidence on the generalization of the results, and the applicability of the proposed approach to other domains.

We also show, as would be expected, that results improve with a larger pool size. That is, the larger the historic of models, the more models there are to select from, and hence the likelihood of retrieving better suited models increases. More interestingly, we show that using the reduced set of features leads to similar or even better results than using the full set. As shown in Table 2, the correlation between the accuracy of the retrieved models and that of the validation models (which represent the new models that would have to be trained), is generally high. The average distance of the accuracy between each pair of models is also generally low.

While these results vary according to the datasets, i.e., there are cases in which the distance between meta-feature vectors is more strongly correlated with model performance, the approach can generally be used to reduce computational resources with the training of ML models, while generally maintaining their predictive performance.

In future work we will explore additional meta-feature optimization techniques, such as dimensionality reduction with PCA and LDA, as well as heuristics such as genetic algorithms and simulated annealing. These techniques can provide a better understanding of the most influential meta-features, further optimizing computational efficiency and prediction accuracy.

Acknowledgments. This work has been supported by the European Union under the Next Generation EU, through a grant of the Portuguese Republic's Recovery and Resilience Plan (PRR) Partnership Agreement, within the scope of the project PRO-DUTECH R3 - "Agenda Mobilizadora da Fileira das Tecnologias de Produção para a Reindustrialização", Total project investment: 166.988.013,71 Euros; Total Grant: 97.111.730,27 Euros.

References

1. Alcobaça, E., Siqueira, F., Rivolli, A., Garcia, L.P.F., Oliva, J.T., de Carvalho, A.C.P.L.F.: MFE: towards reproducible meta-feature extraction. J. Mach. Learn. Res. **21**(111), 1–5 (2020). http://jmlr.org/papers/v21/19-348.html
2. Bol'on-Canedo, V., Mor'an-Fern'andez, L., Cancela, B., Alonso-Betanzos, A.: A review of green artificial intelligence: towards a more sustainable future. Neurocomputing **599**, 128096 (2024)

3. Hoens, T.R., Polikar, R., Chawla, N.V.: Learning from streaming data with concept drift and imbalance: an overview. Progr. Artif. Intell. **1**, 89–101 (2012)
4. Khedher, L., Illán, I.A., Górriz, J.M., Ramírez, J., Brahim, A., Meyer-Baese, A.: Independent component analysis-support vector machine-based computer-aided diagnosis system for Alzheimer's with visual support. Int. J. Neural Syst. **27**(03), 1650050 (2017)
5. Kreuzberger, D., Kühl, N., Hirschl, S.: Machine learning operations (MLOps): overview, definition, and architecture. IEEE Access **11**, 31866–31879 (2023)
6. Liang, T., Glossner, J., Wang, L., Shi, S., Zhang, X.: Pruning and quantization for deep neural network acceleration: a survey. Neurocomputing **461**, 370–403 (2021)
7. Maharana, K., Mondal, S., Nemade, B.: A review: data pre-processing and data augmentation techniques. Glob. Transit. Proc. **3**(1), 91–99 (2022)
8. Rivolli, A., Garcia, L.P., Soares, C., Vanschoren, J., de Carvalho, A.C.: Characterizing classification datasets: a study of meta-features for meta-learning. arXiv preprint arXiv:1808.10406 (2018)
9. Roth, W., et al.: Resource-efficient neural networks for embedded systems. J. Mach. Learn. Res. **25**(50), 1–51 (2024)

Addressing the Limitations of LIME for Explainable AI in Manufacturing: A Case Study in Textile Defect Detection

João Pereira[2], Filipe Oliveira[1], Miguel Guimarães[1], Davide Carneiro[1,2(✉)], Miguel Ribeiro[3], and Gilberto Loureiro[3]

[1] INESC TEC, R. Dr. Roberto Frias, 4200 Porto, Portugal
davide.r.carneiro@inesctec.pt

[2] Escola Superior de Tecnologia e Gestão, Instituto Politécnico do Porto, Felgueiras, Portugal

[3] Smartex, Rua Manuel Pinto de Azevedo 567 Armazém 1, 4100-320 Porto, Portugal

Abstract. Explainable Artificial Intelligence (xAI) techniques are nowadays widely accepted as one of the paths towards addressing the interpretability and transparency issues of using black box models. Such techniques may allow to understand, to a certain extent, how or why a model produced a certain output, which may even help identify problems with the model or the data. As in many other domains, the use of xAI techniques in the context of manufacturing is seen as fundamental towards understanding model outputs, supporting informed decision-making, or enabling more human-centric approaches. In this paper, we specifically look at LIME, one of the most widely used approaches to xAI, and at how it needs to be adapted to the manufacturing context. Specifically, we show how the image permutations introduced by LIME might deceive the underlying model and generate poor explanations, and propose a methodology to address this issue. The specific use-case is on defect detection in the textile manufacturing industry.

Keywords: Explainable AI · Explainability · Defect Detection · Manufacturing

1 Introduction

Over the past decade, the development of Artificial Intelligence (AI) was unprecedented, and is now intertwined in virtually every aspect of our lives. Developments in new model architectures, new algorithms and new hardware allowed AI to take big leaps in performance, especially with the advent of Deep Learning.

However, this development was much more evident in the technical aspects than in other secondary but equally relevant dimensions. Indeed, while the surprising abilities of AI took the world by surprise and have quickly been integrated into our day-to-day living, fundamental dimensions such as the legal or ethical

© The Author(s) 2025
K. Alexopoulos et al. (Eds.): ESAIM 2024, LNME, pp. 262–270, 2025.
https://doi.org/10.1007/978-3-031-86489-6_27

frameworks of its use are still being discussed and set up, and apparently with far more difficulty than AI's technical aspects [2,6,8].

Aspects such as explainability and interpretability of AI models are part of the desiderata, with the goal of making these models more transparent and open to human interpretation, and ultimately allowing for more oversight, control, alignment and compliance. In that regard, many tools, techniques and frameworks are now being put forward that allow, to some extent, to peer into the internal processes of AI or to generate post-hoc information that improves human-decision processes within an AI ecosystem [3,4].

Existing tools are, however, generic in nature and are often inadequate to be directly used in contexts with specific requirements or characteristics, such as the manufacturing one. This is the case of approaches such as SHAP (SHapley Additive exPlanations) or LIME (Local Interpretable Model-agnostic Explanations), when used with image classification tasks, which rely on perturbing superpixels of the images by painting them with a black background [5]. This may mislead the underlying model, and limit the quality and faithfulness of the explanation.

This paper addresses the causes for this issue, and proposes a methodology to address it. We validate this methodology in the specific domain of textile defect detection, with a model trained on the MVTec dataset [1]. While we develop the case study using LIME, this contribution extends to other similar approaches such as SHAP.

2   Contextualization

LIME is a technique designed to explain predictions of ML models, including those used in computer vision tasks. The core idea behind LIME is to provide explanations that are both interpretable and locally faithful to the model's behavior around the specific prediction being explained, which entails several steps.

First, for any given image to be explained, LIME generates several perturbed versions. For object detection tasks, as is the case, this often involves modifying random regions of the image. These regions are called super-pixels, which are obtained using a segmentation algorithm. The vanilla version of LIME paints the perturbed superpixels black. This process produces a collection of modified images that differ more or less from the original, depending on the number of superpixels that were perturbed [9].

Next, the modified images are passed through the underlying object detection model being explained, to obtain predictions for each one. This helps understand how different parts of the image influence the model's detection results: if the prediction for two perturbed versions of the image changes significantly when one superpixel was changed, that region is most likely important for the prediction.

A local surrogate interpretable model is then trained, such as a linear regression, to approximate the behavior of the complex model in the local region defined by the perturbed images. The simple model aims to capture the relationships between features (the perturbed super-pixels) and the prediction output.

One can then look at the relevance of each feature in the surrogate model, to identify the super-pixels that are relevant in the original image to classify the object.

One key aspect in this process is on how to fill in the perturbed sections of the images that are generated. The vanilla version of LIME fills them in black. However, while this tends to work generally well in generic tasks, there are cases in which this black artifact that is created in the image might be confused by the underlying model with an actual object. This is exemplified in Fig. 1. The top row shows three detections of the defect *color*, which may be caused by oil or paint stains in the fabric. The bottom row shows three examples of perturbations generated during the use of LIME to explain the predictions, using different backgrounds, and the output of the underlying model for each one.

In d), the actual defect is not detected (false negative), and the model detects three *color* defects in the super-pixels that were painted black by LIME in this perturbation (false positives). Using other colors or even more complex patterns might have similar results. In e) and f) a complex background and the color pink were respectively used instead of black. In the case of e) the actual defect, while visible, is not detected (as in d)), while two of the superpixels are mistakenly classified by the model as *cut* and *metal contamination*, which are another two defects that the model has been trained on. In f) the model wrongfully detects three color defects on super-pixels that were painted pink in the perturbation, when actually no defect should have been detected since the actual defect is hidden by the perturbation.

Thus, when using LIME, it is fundamental that an appropriate color or background is selected for filling in the perturbed superpixels, so that these are not mistaken by actual objects by the underlying model. While this may be laborious to achieve on a case-by-case basis, in Sect. 3 we propose a methodology to do this automatically for any use-case.

Another relevant aspect is that the quality of the explanations generated by LIME is directly dependent on the quality of the underlying model, and on the eventual similarity between the objects and the perturbations created by LIME. In this specific use-case, we are using a previously trained model of the YOLOv7 class to detect 4 different defects (color, cut, hole and metal contamination). To train the model, we used the MVTec Anomaly Detection Dataset [1].

3   Methodology

This section proposes a methodology to address one of the key limitations of LIME, described in the previous section and made evident in Fig. 1. We aim to automate the process of selecting a suitable color or pattern for LIME to generate perturbations for each specific use-case, ensuring that the underlying model, whatever it is, does not mistake perturbed super-pixels with the objects/defects being detected. Doing this by hand may be time-consuming and not exhaustive, so automating the whole process is advantageous for the manufacturing domain.

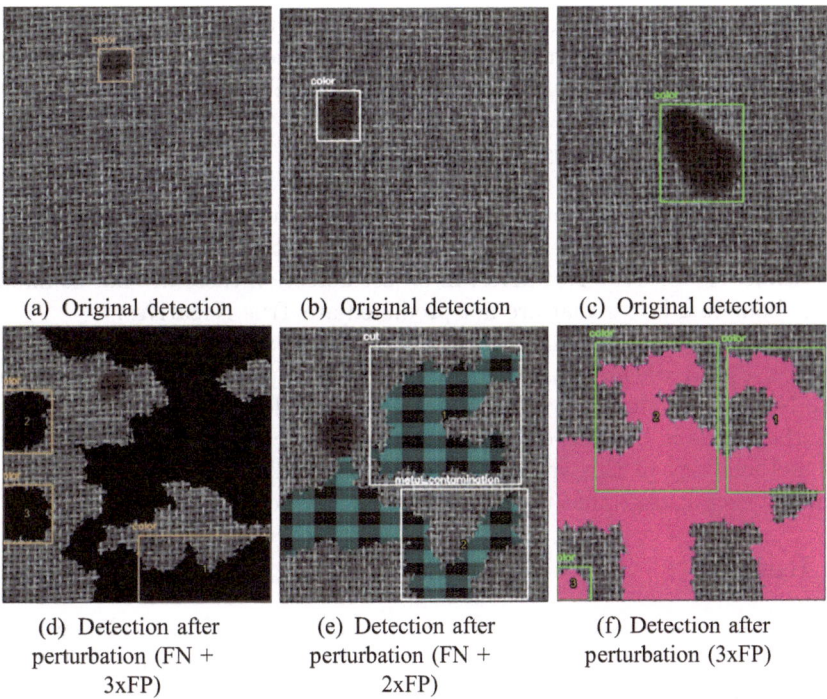

(a) Original detection (b) Original detection (c) Original detection

(d) Detection after
perturbation (FN +
3xFP)

(e) Detection after
perturbation (FN +
2xFP)

(f) Detection after
perturbation (3xFP)

Fig. 1. Top row shows textile defect detections by the underlying YOLOv7 model. Bottom row shows cases in which the perturbations are mistaken by actual defects (FP: False Positive) and/or in which the actual defect is no longer detected (FN: False Negative).

The proposed methodology takes as input a group of images with defects and their corresponding bounding boxes (ground truth), which contain the coordinates and class of each defect. For the sake of completeness, these images should contain a few examples of each class being detected. It also takes as input a group of colors and/or background patterns that the user intends to test as potential candidates to fill-in the perturbed superpixels.

For each input image and each background being tested, we then generate a pre-determined number of copies with random perturbations. The underlying model being explained is then used to predict on each of the perturbed images and its predictions will be compared with those of the same model on the original image. The general intuition is that if the predictions of the model when the defect is visible are generally the same as those on the original image, then the model is not significantly affected by the background color/pattern being tested. On the other hand, when the defect is not visible, the model should not detect any defects. If it does, it is likely confusing the perturbed superpixels with objects in the image that are not there (false positives).

The model is thus asked to predict on each perturbed image, and the performance of each background color/pattern is assessed for each individual class. Namely, we consider the four following cases: True Positive: the model detected a defect in the perturbed version and the original image did indeed have the same defect in the same region; False Positive: the model detected a defect in the perturbed version but there wasn't a defect in that region of the original image or it was of a different class; and False Negative: the model did not detect a defect in the perturbed version but the defect (or the majority of it) was actually visible. As for True Negatives, in object detection these are not counted as all the bounding boxes that are not predicted are True Negatives.

It must also be noted that multiple cases may exist in the same image. For instance, Figs. 1 d) and e) show examples of multiple False Positives and one False Negative in the same image, as the original defect is not detected in the perturbed version, but non-existing defects are detected. The results of applying this methodology in this specific use-case are detailed in Sect. 4, in which the results of each background tested are reported by defect.

4   Results

This section presents the results of evaluating the proposed methodology. In this use case, in the domain of textiles defect detection, we considered 9 different colors and patterns to create the perturbations in LIME, and use the proposed methodology to find the most suitable background for this specific domain. The colors and backgrounds were defined arbitrarily since, due to the nature of Deep Learning models, it is virtually impossible to estimate which will not be confused by the model with the defects being detected.

We evaluate the suitability of each background through the ability of the baseline model to identify defects in the perturbed images, using the metrics of the PASCAL VOC object detection challenge, which are based on the Precision x Recall curve and Average Precision [7]. In the computation of the metrics we used an $IoU = 0.5$.

In order to compute these metrics, it is also necessary to define the threshold at which a defect is considered visible in the perturbed images. Indeed, when LIME creates the perturbed images, three different cases can occur: the defect is completely visible, the defect is partially hidden by one or more perturbations, or the defect is completely hidden. Images that are partially hidden must still be labeled as containing the defect or not, so that the model can then be accurately evaluated. For the results reported herein, we assumed that a defect is only considered visible if 10% or less of the area of its bounding box is covered by the perturbation.

Table 1 shows the counts of the visible defects in the 2000 perturbations generated for each background (18.000 in total) in the *Ground Truth* column, and the detections by the underlying model in the *Detections* column. It is clearly visible that the model tends to mistake certain backgrounds with actual defects, as happens with the *color* defect in the first two colors and in patterns 1 and 3, or in the *cut* defect in pattern 4.

Table 1. Number of defects of each type visible in the 2000 images with perturbations (ground truth) vs. the number of defects detected by the model (detections), by background.

Background	Ground Truth				Detections			
	color	cut	hole	metal	color	cut	hole	metal
(0,0,0)	146	52	53	134	9136	127	22	53
(255,15,192)	116	33	71	142	10449	0	9	10
(255,255,255)	123	37	52	139	73	3672	1041	383
(0,177,64)	125	43	55	152	155	29	180	173
Pattern 1	108	39	49	171	8193	50	27	54
Pattern 2	108	44	58	142	232	127	249	227
Pattern 3	139	41	57	143	6387	0	14	20
Pattern 4	101	39	64	148	86	6083	259	0
Pattern 5	118	33	59	147	11	4031	415	5137

Table 2 summarizes the results for the different backgrounds compared. Specifically, it shows the average precision by class, and the mean average precision (mAP). It allows to compare the different backgrounds in terms of how they confound the model, and also how the same background is more or less confused with each of the defects.

For this specific use case, the best background is Pattern 2, with an mAP = 0.62, followed by color R = 255, G = 15, B = 192, with an mAP = 0.43. All the

Table 2. Performance of the model when predicting on images with permutations generated using different colored/patterned backgrounds. The last row shows the performance of the model when predicting on the original images.

Background	AP per class				mAP
	color	cut	hole	metal	
(0,0,0)	0.0	0.08	0.15	0.33	0.14
(255,15,192)	0.50	0.15	0.41	0.64	0.43
(255,255,255)	0.0	0.0	0.11	0.07	0.05
(0,177,64)	0.43	0.0	0.03	0.14	0.15
Pattern 1	0.0	0.01	0.10	0.14	0.06
Pattern 2	**0.63**	**0.51**	**0.64**	**0.70**	**0.62**
Pattern 3	0.0	0.0	0.11	0.14	0.06
Pattern 4	0.38	0.0	0.04	0.30	0.18
Pattern 5	0.09	0.0	0.04	0.06	0.05
Baseline	1.0	1.0	1.0	0.8	0.95

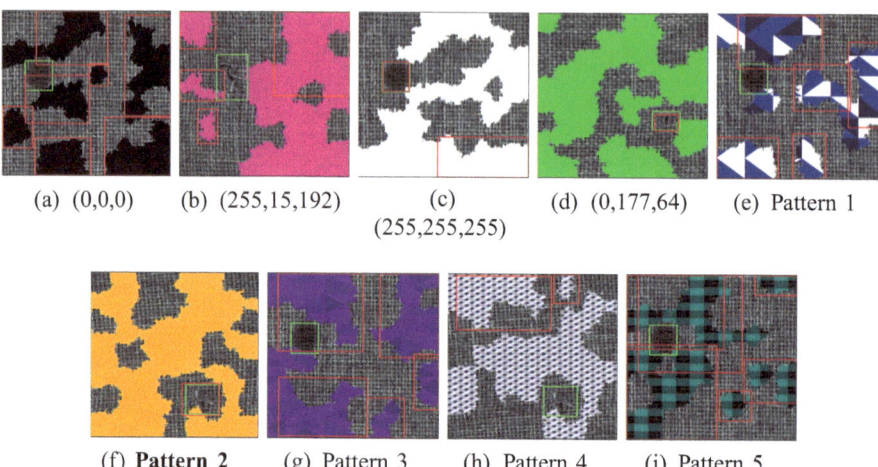

Fig. 2. Sample (out of 18.000) of one randomly selected perturbation generated by LIME for each background, ground truth (green), and detections by the underlying model (red).

remaining backgrounds have very poor average results, despite some acceptable results for some specific classes.

Figure 2 shows the same image, with a color defect, and one randomly selected perturbation for each background. The ground truth is shown in green (when visible) and the detections of the underlying model are shown in red. Pattern 2, depicted in (f), was the background that achieved the best results for this specific use-case, out of the 9 tested.

5   Discussion and Conclusions

One key contribution of this paper is to show that explainability methods based on image permutations may be inadequate for the manufacturing domain, as the black superpixels introduced in the images are often mistaken by the underlying model and detected as defects. This negatively impacts the quality of the generated explanations.

Intuition could lead to assume that choosing a background that is not visually similar to the defects being detected would solve this problem. The second contribution of this paper is to show that this is not the case, as Figs. 1 and 2 show. This is due to the fact that Deep Learning models construct unique internal representations of the objects, which may be very different from our own representations. This work also shows how the performance of Deep Learning models can change dramatically and unpredictably when new artifacts are introduced in the images. This lack of transparency and predictability highlights the need for approaches such as the proposed one.

Finally, it should be highlighted that the main contribution of this paper is not the identification of the best background for this particular use-case, but rather the methodology followed to do it, since each problem and model is different. While some particular backgrounds may be generally better across different problems/domains, this remains to be studied.

Current work covers two aspects. On the one hand, we are developing objective metrics for evaluating the quality of explanations. This will allow to assess two different aspects separately: the quality of the model and the quality of the generated explanations. On the other hand, we are including permuted images together with the original training data when training models, so that they naturally learn to ignore the permuted superpixels. The methodology proposed in this paper is used to decide which background to use in each problem prior to the training, minimizing the necessary training data.

The approach described in this paper is thus one step forward towards the inclusion of more human-centric and transparent approaches in industrial AI applications, in which human oversight for end-users and developers is seen as fundamental.

Acknowledgments. This work has been supported by the European Union under the Next Generation EU, through a grant of the Portuguese Republic's Recovery and Resilience Plan (PRR) Partnership Agreement, within the scope of the project PRODUTECH R3 - "Agenda Mobilizadora da Fileira das Tecnologias de Produção para a Reindustrialização", Total project investment: 166.988.013,71 Euros; Total Grant: 97.111.730,27 Euros.

References

1. Bergmann, P., Batzner, K., Fauser, M., Sattlegger, D., Steger, C.: The mvtec anomaly detection dataset: a comprehensive real-world dataset for unsupervised anomaly detection. Int. J. Comput. Vision **129**(4), 1038–1059 (2021)
2. Coeckelbergh, M.: Artificial intelligence: some ethical issues and regulatory challenges. Technology and regulation **2019**, 31–34 (2019)
3. Ding, W., Abdel-Basset, M., Hawash, H., Ali, A.M.: Explainability of artificial intelligence methods, applications and challenges: a comprehensive survey. Inf. Sci. **615**, 238–292 (2022)
4. Ehsan, U., Liao, Q.V., Muller, M., Riedl, M.O., Weisz, J.D.: Expanding explainability: Towards social transparency in ai systems. In: Proceedings of the 2021 CHI conference on human factors in computing systems. pp. 1–19 (2021)
5. Garreau, D., Luxburg, U.: Explaining the explainer: a first theoretical analysis of LIME. In: Proceedings of the Twenty Third International Conference on Artificial Intelligence and Statistics, pp. 1287–1296. PMLR, June 2020
6. Naik, N., Hameed, B., Shetty, D.K., Swain, D., Shah, M., Paul, R., Aggarwal, K., Ibrahim, S., Patil, V., Smriti, K., et al.: Legal and ethical consideration in artificial intelligence in healthcare: who takes responsibility? Front. Surg. **9**, 266 (2022)
7. Padilla, R., Netto, S.L., Da Silva, E.A.: A survey on performance metrics for object-detection algorithms. In: 2020 International Conference on Systems, Signals and Image Processing (IWSSIP), pp. 237–242. IEEE (2020)

8. Stahl, B.C., Stahl, B.C.: Ethical issues of ai. Artificial Intelligence for a better future: an ecosystem perspective on the ethics of AI and emerging digital technologies, pp. 35–53 (2021)

9. Tan, L., Huang, C., Yao, X.: A concept-based local interpretable model-agnostic explanation approach for deep neural networks in image classification. In: International Conference on Intelligent Information Processing, pp. 119–133. Springer (2024)

Multifaceted Applications of Federated Learning: Beyond Neural Networks

Tatjana Legler[1,2]($\boxtimes$), Vinit Hegiste[1], and Martin Ruskowski[1,2]

[1] Chair of Machine Tools and Control Systems, University of Kaiserslautern-Landau (RPTU), 67663 Kaiserslautern, Germany
tatjana.legler@rptu.de
[2] Innovative Factory Systems, German Research Center for Artificial Intelligence (DFKI), 67663 Kaiserslautern, Germany

Abstract. The advent of Federated Learning (FL) has brought about a revolutionary change in the field of machine learning, enabling the decentralised training of models across a multitude of devices while simultaneously maintaining the confidentiality of the data. In contrast to conventional centralized methodologies, FL maintains the localisation of data, with only model updates being shared. This methodology enhances model generalisation and stability without compromising data sovereignty. A variety of machine learning techniques, including support vector machines (SVMs) and decision trees, can be effectively utilised within the context of FL frameworks. SVMs offer efficient solutions for classification tasks with minimal computational overhead, while decision trees provide interpretable models for both classification and regression. This paper explores the application of these methods in FL settings, highlighting their advantages and potential use cases in diverse industries, particularly in manufacturing. Furthermore, it discusses the integration of reinforcement learning with FL, emphasising its potential for enhancing intelligent and adaptable decentralised systems.

Keywords: Federated Learning · Machine Learning Algorithms · Privacy-preserving Techniques

1 Introduction

Federated Learning (FL) represents a paradigm shift in machine learning by facilitating a decentralized learning approach [21]. Unlike traditional centralized learning models, where data is consolidated on a single server for training [6,15], FL enables model training across multiple decentralized devices or nodes while keeping all the training data localized. One of the primary objectives of FL is to utilize a broad, heterogeneous database without the need to share sensitive or proprietary information [16]. This method ensures that only model updates, such as network weights, are shared between nodes, rather than raw data like sensor data or images [12]. Such an approach preserves data privacy and benefits all participants by combining the advantages of data sovereignty with the

© The Author(s) 2025
K. Alexopoulos et al. (Eds.): ESAIM 2024, LNME, pp. 271–278, 2025.
https://doi.org/10.1007/978-3-031-86489-6_28

collaborative power of model training, thereby enhancing model generalizability and stability [30].

The introduction of Federated Averaging (FedAvg) by McMahan et al. [21] established a structured approach for applying FL through an iterative process. Initially, a central server initializes a global model and distributes it to all participating devices or nodes. Each node then independently trains the model on its own local dataset. After local training, nodes send their model updates, such as weights, back to the central server. The server aggregates these updates, typically by averaging, to update the global model. The updated global model is then sent back to the nodes. This process, known as a communication round, repeats until the model reaches the desired level of accuracy or meets other convergence criteria (see Fig. 1). This iterative cycle leverages distributed data sources while maintaining the privacy and security of the data handled by each node.

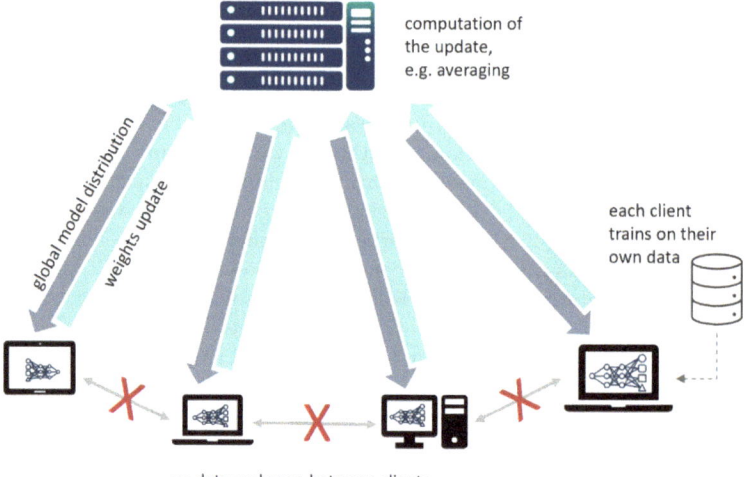

Fig. 1. This illustration depicts the federated learning process where a global model is distributed to multiple clients. Each client independently trains the model on their local data without exchanging data with other clients. The locally trained models' updates are sent back to a central server, which computes the aggregated update, typically through averaging. [18]

2   Machine Learning Methods in Federated Settings

While neural networks are known for their ability to handle complex and high-dimensional data, they are not always the most efficient or necessary approach for every problem. This is particularly true in FL settings, where the primary objective is to perform decentralized machine learning effectively and efficiently.

Neural networks, especially deep learning models, are resource-intensive in terms of both computational power and data requirements [9]. Their high dimensionality increases communication overhead, potentially leading to inefficiencies and delays in the FL process [32].

Moreover, many real-world problems do not require the extensive modeling power of neural networks. Simpler approaches, such as Support Vector Machines (SVMs) or tree-based models, can often achieve comparable or sufficient accuracy with significantly less computational overhead and complexity [2]. SVMs are particularly well-suited for scenarios where the margin of separation and generalization to unseen data are more critical than capturing complex data patterns.

2.1 Support Vector Machines

SVM are a class of supervised learning algorithms used for classification and regression tasks. They are based on the concept of finding a hyperplane that best separates the data points of different classes in a high-dimensional space [26]. The main idea is to maximize the margin, which is the distance between the hyperplane and the closest data points from each class, known as support vectors. This margin maximization leads to better generalization and robustness of the classifier. SVMs can handle both linear and non-linear classification by using kernel functions, such as polynomial, radial basis, and sigmoid function, to map the input data into higher-dimensional spaces where it becomes linearly separable [17]. The theoretical foundations of SVMs are rooted in statistical learning theory, particularly in the concept of Structural Risk Minimization, which aims to find a balance between model complexity and fitting the training data to minimize the generalization error [29]. In a federated setting, the solution calculated by SVM must accommodate not only one dataset, but multiple subsets as accessible per client. This may result in a slightly different angle of the hyperplane, as illustrated in Fig. 2.

The applications of SVM in manufacturing are numerous and diverse. The following are a few illustrative examples: In [1], SVMs are used to diagnose mechanical faults in motors by analyzing vibration signals collected from accelerometers. The SVMs classify the transformed vibration data to identify specific issues such as unbalance, misalignment, and mechanical looseness. In [28], SVMs predict anomalies within the manufacturing process. Upon detecting an anomaly, the system dynamically reconfigures itself to mitigate the issue, rerouting jobs to different machines or adjusting processing paths to maintain optimal operation and prevent overloading any single machine. In [19], SVMs are employed to predict manufacturing lead times. The model categorizes the total manufacturing time of products into different duration ranges, using production and work order data that undergo preprocessing and feature selection to enhance model performance.

A summary of the applications of SVM as referenced in the literature is presented herewith [22,25]. By analyzing historical data, SVM models can identify the relationships between various input parameters and quality metrics, thereby

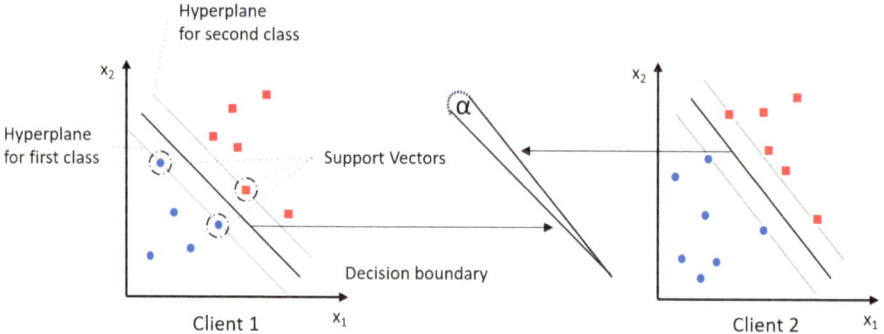

Fig. 2. Illustration of SVM Classifiers. The subfigures show a linear SVM classifier with a straight decision boundary and support vectors along the margins. The SVM aims to maximize the margin between the two classes, represented by red squares and blue circles.

enabling the prediction of product quality during production. By accurately categorizing products, SVMs assist in maintaining consistent quality standards and in the identification of batches that meet or fail to meet the required specifications. SVMs are also employed for the optimization of manufacturing processes, whereby the optimal settings for various machine parameters that result in the highest product quality are identified. Fault diagnosis, as previously mentioned, is also a key application area. By classifying the condition of machines as normal or faulty, SVMs enable the implementation of predictive maintenance strategies, which serve to reduce downtime and prevent catastrophic failures. This process entails training SVM models on historical data to ascertain the impact of varying parameter settings on the output quality. Manufacturers can subsequently utilize these models to simulate diverse scenarios and identify the optimal combination of parameters. To illustrate, in a machining process, SVMs can optimize cutting speed, feed rate, and tool geometry to minimize surface roughness and maximize material removal rate. This optimization not only enhances product quality but also improves process efficiency and reduces costs.

The aforementioned applications can be addressed and extended with FL, as the benefits are consistent across the board: It enhances SVM-based solutions by improving generalization by aggregating model updates from different units, creating a model that better predicts faults across different machines and environments. It enables scalable training across distributed locations, efficiently handling large datasets and allowing local models to quickly detect faults and take corrective action, reducing latency compared to centralized approaches. In addition, FL facilitates continuous learning, where local models are updated with new vibration data and these updates are periodically aggregated to refine the global model, ensuring that the system is always up-to-date with the latest patterns.

2.2 Decision Trees and Random Forrest

A Decision Tree (DT) is a widely used method in machine learning and data analysis for classification and regression tasks [5]. It operates by recursively splitting a dataset into subsets based on the most significant attribute, creating a tree-like structure where each internal node represents a test on an attribute, each branch corresponds to the test's outcome, and each leaf node signifies a final decision or prediction [24]. The selection of the best attribute to split on is typically determined by metrics such as Gini impurity, entropy, or information gain. They have also been employed to summarize associative classification rules, providing a more readable and compact classification mode [4]. They are valued for their simplicity, interpretability, and ability to handle both numerical and categorical data. However, they are prone to overfitting and can be unstable with small data variations. When employed in an ensemble, as is the case with random forest, the disadvantage of overfitting is mitigated [31].

In addition to similar areas of application as mentioned for SVM [23], DT can be used for decision support, e.g. for spare part configuration [3] as they are additionally better to interpret. [11] trained an incremental decision tree in a federated manner. For this they utilize 'Very Fast Decision Tree' (VFDT) as proposed in [7] and trained it in a vertical FL setting. [27] proposed a subtree-based horizontal FL method that accelerates model convergence and reduces communication costs while maintaining accuracy. Their FS-Boost approach learns one level of the tree at a time.

2.3 Reinforcement Learning

In distributed and decentralized systems, FL combined with reinforcement learning (FDRL) leverages the strengths of both methodologies to tackle complex, distributed learning tasks, such as in robotics. This powerful combination ensures data privacy while enabling seamless knowledge transfer between entities, thereby enhancing overall system intelligence and adaptability. FDRL frameworks, like FDRL, facilitate secure information sharing to build high-quality models while maintaining privacy protections [33]. Techniques such as reward shaping improve training efficiency and policy quality without compromising client confidentiality [14]. Furthermore, frameworks like Lifelong Federated Reinforcement Learning (LFRL) enhance robot navigation by fusing and transferring prior knowledge, allowing robots to quickly adapt to new environments [20]. This collaborative approach also supports multi-robot systems, enabling efficient task scheduling and improved learning performance across various applications [8,13]. In a modular factory setup, where each robot module is independent, as exemplified by the configuration depicted in [10], each module can be regarded as a FL client and train on their own objects when separated. However, upon connection, they share their updates and form a new global model. The potential for transferring learned skills across different entities using FDRL remains an open question, highlighting the need for further research in this area.

3   Conclusion and Outlook

This paper demonstrates the extensive potential applications of Federated Learning (FL) in manufacturing, with significant opportunities yet to be fully explored. Our exploration highlights numerous applications for Support Vector Machines (SVM) in areas such as defect detection, quality prediction, and parameter optimization. Additionally, Federated Distributed Reinforcement Learning shows considerable promise within robotics, particularly for path planning and grasp optimization. Our findings reveal that FL extends beyond neural networks, effectively incorporating a variety of machine learning methods. By combining different machine learning techniques with FL, robust, scalable, and privacy-preserving data analysis can be achieved. Each method offers specific strengths and alternatives that are computationally less demanding and more suited to particular types of data or tasks. The increasing awareness and concern over personal data usage, coupled with the enforcement of stringent data protection laws worldwide, underscore the necessity of privacy-preserving methods. By adhering to principles of data minimization and locality, FL aligns well with regulatory requirements and public sentiment, making it an indispensable tool in our data-driven world.

Future work will build on these findings, focusing on the combination of FL and SVM for planning optimization, and FL and Reinforcement Learning for robotics applications such as path planning and grasp optimization. This study underscores the versatility and potential of FL across a wide array of applications, paving the way for its broader adoption and integration into modern manufacturing environments. The continuous evolution of FL methodologies promises to further enhance data-driven decision-making processes while ensuring privacy and regulatory compliance.

Acknowledgment. This work was funded by the Carl Zeiss Stiftung, Germany under the Sustainable Embedded AI project (P2021-02-009).

References

1. Baccarini, L.M.R., Rocha e Silva, V.V., de Menezes, B.R., Caminhas, W.M.: Svm practical industrial application for mechanical faults diagnostic. Expert Syst. Appl.**38**(6), 6980–6984 (2011)
2. Balabin, R.M., Lomakina, E.I.: Support vector machine regression - an alternative to neural networks for analytical chemistry? comparison of nonlinear methods on near infrared spectroscopy data. Analyst **136**(8), 1703–1712 (2011)
3. Cantini, A., Peron, M., de Carlo, F., Sgarbossa, F.: A decision support system for configuring spare parts supply chains considering different manufacturing technologies. Int. J. Prod. Res. **62**(8), 3023–3043 (2024)
4. Chen, Y.L., Hung, L.T.H.: Using decision trees to summarize associative classification rules. Expert Syst. Appl. **36**(2), 2338–2351 (2009)
5. Costa, V.G., Pedreira, C.E.: Recent advances in decision trees: an updated survey. Artif. Intell. Rev. **56**(5), 4765–4800 (2023)

6. Dean, J., et al.: Large scale distributed deep networks. In: Pereira, F., Burges, C.J., Bottou, L., Weinberger, K.Q. (eds.) Advances in Neural Information Processing Systems, vol. 25. Curran Associates, Inc. (2012)

7. Domingos, P., Hulten, G.: Mining high-speed data streams. In: Ramakrishnan, R., Stolfo, S., Bayardo, R., Parsa, I. (eds.) Proceedings of the Sixth ACM SIGKDD International Conference on Knowledge Discovery and Data Mining, pp. 71–80. ACM, New York, NY, USA (2000)

8. Feng, W., Liu, H., Peng, X.: Federated reinforcement learning for sharing experiences between multiple workers. In: 2023 International Conference on Machine Learning and Cybernetics (ICMLC), pp. 440–445. IEEE (2023)

9. Frey, N.C., et al.: Benchmarking resource usage for efficient distributed deep learning. In: 2022 IEEE High Performance Extreme Computing Conference (HPEC), pp. 1–8. IEEE (2022)

10. Gafur, N., Kanagalingam, G., Wagner, A., Ruskowski, M.: Dynamic collision and deadlock avoidance for multiple robotic manipulators. IEEE Access 10, 55766–55781 (2022)

11. Han, Z., Ge, C., Wu, B., Liu, Z.: Lightweight privacy-preserving federated incremental decision trees. IEEE Trans. Serv. Comput. 9, 1–13 (2022)

12. Hegiste, V., Legler, T., Ruskowski, M.: Application of federated machine learning in manufacturing. In: 2022 International Conference on Industry 4.0 Technology (I4Tech), pp. 1–8. IEEE (2022)

13. Ho, T.M., Nguyen, K.K., Cheriet, M.: Federated deep reinforcement learning for task scheduling in heterogeneous autonomous robotic system. IEEE Trans. Autom. Sci. Eng. 21(1), 528–540 (2024)

14. Hu, Y., Hua, Y., Liu, W., Zhu, J.: Reward shaping based federated reinforcement learning. IEEE Access 9, 67259–67267 (2021)

15. Iandola, F.N., Moskewicz, M.W., Ashraf, K., Keutzer, K.: Firecaffe: near-linear acceleration of deep neural network training on compute clusters. In: 2016 IEEE Conference on Computer Vision and Pattern Recognition (CVPR). IEEE (2016)

16. Kairouz, P., McMahan, H.B., Avent, B., Bellet, A., Bennis, M., Nitin Bhagoji, A., et al.: Advances and open problems in federated learning. FNT in Machine Learning (Foundations and Trends in Machine Learning) 14(1–2), 1–210 (2021)

17. Kurani, A., Doshi, P., Vakharia, A., Shah, M.: A comprehensive comparative study of artificial neural network (ann) and support vector machines (svm) on stock forecasting. Annal. Data Sci. 10(1), 183–208 (2023)

18. Legler, T., Hegiste, V., Ruskowski, M.: Mapping of newcomer clients in federated learning based on activation strength. In: 32nd International Conference Flexible Automation and Intelligent Manufacturing (2023)

19. Lim, Z.H., Yusof, U.K., Shamsudin, H.: Manufacturing lead time classification using support vector machine. In: Badioze Zaman, H., Smeaton, A.F., Shih, T.K., Velastin, S., Terutoshi, T., Mohamad Ali, N., Ahmad, M.N. (eds.) IVIC 2019. LNCS, vol. 11870, pp. 268–278. Springer, Cham (2019). https://doi.org/10.1007/978-3-030-34032-2_25

20. Liu, B., Wang, L., Liu, M.: Lifelong federated reinforcement learning: a learning architecture for navigation in cloud robotic systems. IEEE Robot. Automation Lett. 4(4), 4555–4562 (2019)

21. McMahan, B., Moore, E., Ramage, D., Hampson, S., Arcas, B.A.y.: Communication-efficient learning of deep networks from decentralized data. In: Singh, A., Zhu, J. (eds.) Proceedings of the 20th International Conference on Artificial Intelligence and Statistics. Proceedings of Machine Learning Research, vol. 54, pp. 1273–1282. PMLR, Fort Lauderdale, FL, USA (2017)

22. Paturi, U.M.R., Cheruku, S.: Application and performance of machine learning techniques in manufacturing sector from the past two decades: a review. Materials Today: Proc. **38**, 2392–2401 (2021)
23. Plathottam, S.J., Rzonca, A., Lakhnori, R., Iloeje, C.O.: A review of artificial intelligence applications in manufacturing operations. J. Adv. Manuf. Process. **5**(3) (2023)
24. Rokach, L., Maimon, O.: Decision trees. In: Maimon, O., Rokach, L., Maimon, O.Z., Roah, L. (eds.) Data Mining and Knowledge Discovery Handbook, pp. 165–192. SpringerLink Bücher, Springer, US, Boston, MA (2005)
25. Rostami, H., Dantan, J.Y., Homri, L.: Review of data mining applications for quality assessment in manufacturing industry: support vector machines. Int. J. Metrol. Quality Eng. **6**(4), 401 (2015)
26. Saravanan, K., Prakash, R., Balakrishnan, C., Kumar, G.V.P., Siva Subramanian, R., Anita, M.: Support vector machines: Unveiling the power and versatility of svms in modern machine learning. In: 2023 3rd International Conference on Innovative Mechanisms for Industry Applications (ICIMIA), pp. 680–687. IEEE (2023)
27. Shimamura, K., Takamaeda-Yamazaki, S.: Fs-boost: Communication-efficient federated subtree-based gradient boosting decision trees. In: 2024 IEEE 21st Consumer Communications & Networking Conference (CCNC). pp. 839–842. IEEE (2024)
28. Shin, H.J., Cho, K.W., Oh, C.H.: Svm-based dynamic reconfiguration cps for manufacturing system in industry 4.0. Wireless Commun. Mob. Comput. **2018**, 1–13 (2018)
29. Vapnik, V.N., Vapnik, V., et al.: Statistical learning theory. Wiley, New York (1998)
30. Zhang, C., Xie, Y., Bai, H., Yu, B., Li, W., Gao, Y.: A survey on federated learning. Knowl.-Based Syst. **216**, 106775 (2021)
31. Zhang, M., Tao, F., Zuo, Y., Xiang, F., Wang, L., Nee, A.: Top ten intelligent algorithms towards smart manufacturing. J. Manuf. Syst. **71**, 158–171 (2023)
32. Zhang, W., Zhou, T., Lu, Q., Yuan, Y., Tolba, A., Said, W.: Fedsl: a communication-efficient federated learning with split layer aggregation. IEEE Internet Things J. **11**(9), 15587–15601 (2024)
33. Zhuo, H.H., Feng, W., Lin, Y., Xu, Q., Yang, Q.: Federated deep reinforcement learning

Ethics, Data and Privacy Management for AI Solutions in Manufacturing at SMEs (AIRISE)

Susan Hommerson[1(✉)], Chiara Gallese[1], Fanny Garel[2], Angela Aleksovska[1], and Harry Bikas[3]

[1] Eindhoven University of Technology, Eindhoven, The Netherlands
s.m.hommerson@tue.nl
[2] Czech University of Technology, Prague, Czech Republic
[3] Laboratory for Manufacturing Systems and Automation, University of Patras, Patras, Greece

Abstract. Ethics has become an increasingly important topic within the development of Artificial Intelligence systems. The anticipated European AI act has a strong foundation in 7 ethical values of the Assessment List for Trustworthy Artificial Intelligence (ALTAI): human agency and oversight, technical robustness and safety, privacy and data governance, transparency, diversity, non-discrimination and fairness, environmental and societal wellbeing, and accountability. In addition, data management and privacy are other important aspects of responsible AI: from development towards implementation. In many research fields, a form of ethical management or review is common practice. However, in the field of AI and manufacturing this is relatively new.

Within the EU project: *Artificial Intelligence in Manufacturing for Sustainability at SMEs* (AIRISE), AI experts will support SMEs active in manufacturing, in the uptake of AI by conducting tailor-made experiments. One requirement for these experiments is the management of ethics, data and privacy. At the same time, AIRISE will also deliver this as a service to SMEs. Ethics management allows for reflection on business processes, safe working and possible improvements.

This work presents our way of service delivery, tutoring, training modules and review procedure, based on the ALTAI principles, GDPR and FAIR principles. Our work lays the foundations for compliance to multiple EU legislations such as the AI act and GDPR, delivered in a user-friendly service. In the nearby future, we anticipate optimizing our ethics, privacy and data management services by developing digital tooling. Furthermore, we will thematically analyze the obtained data and identify which ethical issues are important for SMEs and manufacturing. With this knowledge, we aim to further support manufacturing SMEs in their business.

Keywords: ethics · privacy · data management · manufacturing · AI · SMEs

© The Author(s) 2025
K. Alexopoulos et al. (Eds.): ESAIM 2024, LNME, pp. 279–286, 2025.
https://doi.org/10.1007/978-3-031-86489-6_29

1 Introduction

1.1 Responsible Data Use in AI

In 2020, the EU has developed a European strategy for data in which they aim to create a single market for data that will ensure Europe's global competitiveness and data sovereignty [1]. Ethics in AI, responsible data management are important topics and broadly discussed in this context. Within Horizon Europe projects, the EU commission lays emphasis on responsible data use in the ethics appraisal [2] or additional ethical work packages. Beneficiaries need to comply to relevant legislations, field and codes of conduct.

The European AI-act is one of the first legislations in the world to regulate the use of Artificial Intelligence. The legislation will be stepwise implemented, in the upcoming years from 2024 onwards [3]. The fundament of the legislation is predicated upon 7 ethical principles noted in the Assessment List for Trustworthy Artificial Intelligence (ALTAI) for self-assessment, formulated by the high-level expert group on artificial intelligence [4]. The principles are: human agency and oversight, technical robustness and safety, privacy and data governance, transparency, diversity, non-discrimination and fairness, environmental and societal wellbeing, and accountability. These principles are key in the EU ethics appraisal schemes for AI research [2]. In terms of privacy, the EU has multiple legislations, the most familiar one is the General Data Protection Regulation (GDPR). The GDPR includes rules to pursue the protection of personal data. One of the GDPR requirements when using personal data for developing new technologies is a Data Protection Privacy Assessment (DPIA) for high-risk personal data in which an assessment of all privacy risks and mitigation are formulated. The European union also supports the principles: Findable, Accessible, Interoperable and Reusable (FAIR) as a framework for data management plans [5]. This allows a concrete project plan for the use of data in research projects. These conditions are extremely important for AI development.

1.2 AI in Manufacturing

AI ethics in manufacturing is a relatively new field, but will likely be very important due to the uptake of AI technologies in manufacturing processes. The field is evolving as stakeholders recognize the importance of ethical considerations in AI deployment. Brintrup et al., 2023 [6] produced a list of trustworthy AI, challenges and risks in manufacturing and how they are presented throughout the life-cycle. In addition, manufacturing companies who are providers of high-risk AI systems will need to implement quality management systems and perform appropriate risk management strategies to fulfill the upcoming AI act requirements [7]. Currently AI act standards are being created by standardization body Cen-Cenelec [8], in the meantime implementing AI ethical principles will serve as a strong fundament for the next steps in compliance. Large-scale companies might be better equipped for compliance in comparison to SMEs due to fewer resources and expertise. Therefore, the project Artificial Intelligence in Manufacturing for Sustainability (AIRISE) [9] focusses on aiding SMEs.

1.3 AIRISE Project

In the AIRISE project, a consortium of European partners will provide consultancy to implement AI solutions for SMEs in manufacturing by doing tailor made experiments. These services aim to improve business processes in the field of design and engineering, process monitoring and control, manufacturing operations, production and supply chains. In order to perform these experiments in an ethical responsible way, the management of ethics, data and privacy is key. At the same time, AIRISE will also deliver this as a service to SMEs as the fundament to be compliant with AI-act requirements. In addition, ethics management allows for reflection on business processes, how personal data is being dealt with, safe working and location of data storage.

Therefore, in this paper we aim to present our way of service delivery, tutoring, training modules and review procedure for SMEs within the EU project AIRISE, based on the ALTAI principles, GDPR and FAIR principles for SMEs as a basis for compliance and good responsible data practices. Furthermore, our approach needs to be user-friendly, easy to understand and simple to implement.

2 Method

2.1 The Ethics Data and Privacy Advisory Committee

To ensure compliance with the ethics requirements throughout the project, a governance structure has been put in place: project Ethical & Data privacy Advisory Committee. This committee has been established immediately after the start of the project. Their tasks are twofold: 1) co set-up and or approve the set-up of the ethics manual and 2) ensure compliance of the experiments throughout the project.

The committee comprises of a representative of each project partner, and will be supervised by an expert Ethics Advisory team from the ethics lead project partner. The lead has a team of experts consisting of an: ethics project leader, data steward, attorney

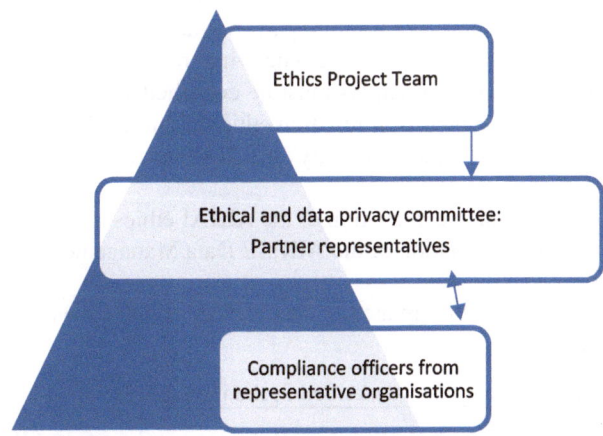

Fig. 1. Governance structure Ethics Data and Privacy Advisory Committee.

in privacy and AI and project assistant. This team took the lead on preparing the material for the procedures on data management, privacy and ethics. They have provided it to the board members for reviewing, technical input and or writing the material. Ultimately, the board approved the final material, and was then named the ethics manual. The members of the board were not necessarily ethics or data experts. They had to be able to review the material or consult an ethics/data expert in their organization for review. The structure of the committee is summarized in Fig. 1.

3   Results

3.1   The Ethics Manual

The Ethics manual identifies the key ethical and legal topics and the procedures for data collection, storage, protection, retention, transfer, destruction, or re-use for the partners and the customers, as well as details on data safety procedures and informed consent. The relevant ethical aspects stated in this document are mainly related to the ethical and safe conduct of AIRISE experiments with participants and the proper use of the collected data. It is intended to be used by the project partners and participating SMEs as a general guideline for AI implementation for all experiments. In addition to the background information, the manual includes work templates for the SMEs to fill in for their experiment. It will be shared with participants taking part in the training sessions. The users of the manual are meant to benefit from clarified procedures for data collection, storage, protection, retention, transfer, destruction, or re-use; details on data safety and informed consent procedures.

Content of the Ethics manual

Chapter 1 introduces the purpose.
Chapter 2 states how to use the manual and the templates in relation to the intended audience of the current document items.
Chapter 3 summarizes the legal and ethical framework used in the creation of the manual: for privacy GDPR and e-Privacy are discussed, ENISA legislation for cybersecurity. Regarding IP rights, it is mandatory for third parties to adhere to contractual agreements. Following the Open Data Directive, all results deriving from EU grants must be published in open access and the FAIR principles must be complied with. Finally, AIRISE will comply with the highest standards of research ethics and integrity, following the European Code of Conduct for Research Integrity (ALLEA) and the European Commission's ethics guidelines.
Chapter 4 includes a detailed explanation of the ALTAI ethics requirements
Chapter 5 provides an explanation of the AIRISE Data Management Plan with leading FAIR principles
Chapter 6 summarizes the rules related to data Privacy and the use of a Privacy Impact Assessment in particular.

Appendices contain the templates described in the manual

- The **Ethics checklist & risk analysis template,** is conceived to be filled in at the start (after 1 month) when the intended use of the experiment is known, halfway, and right before the end of each experiment, before implementation. Not all questions in the ethics checklist are relevant for all experiments; participants should fill in as much as possible, all that is applicable. During the start of the experiment, some questions might be premature; in that case the intention should be filled in, and the document will updated at a later stage during implementation. The checklist comprises of the 7 ALTAI principles, which we aimed to make explainable for manufacturing. Firstly we provide definition of the principle, then the question to be answered and thirdly the explanation of the question, to make it tangible. An example of the principle of Human agency and oversight is given in the Fig. 2.

Definition:
How the AI behaves in the environment and the impact it has on individuals and society overall. The AI should not be free to decide in autonomy. A human being should always be present who understands how the machine works and the consequences of the use of the machine. It should oversee the entire project => human in the loop. It is up to the individual research group how to structure the accountability framework. This can be done manually before starting the machine or during the whole process for example.

Question:
Can you put in place measures to ensure human oversight?

Explanation:
What mechanism are in place to prevent the system to act against the operator's will? Can the operator stop the system in case of emergency? Is there a reset button? Can the human correct the system output?

Fig. 2. Example of Human agency and oversight in the Ethics checklist

In addition, a standard risk analysis needs to be filled for the ethical principle of technical robustness and safety to account for: design and technical faults, possible misuse etc. Moreover, SMEs are welcome to identify other risks they deem relevant in their experiment. For each risk, an appropriate mitigation action needs to be formulated.

– The **Research Data Management (RDM) checklist** is designed to be used before the start of each experiment, in order to check if each item is present.

In case personal research data is required in the experiment, the following also apply

– **Privacy Impact Assessment** should be filled in before the start of the experiment.
– **Informed Consent Form.** In case personal data is collected from participants, this needs to be filled-in with the participant. A copy must be filed and given to the participant of which the data is collected. All information that is relevant to the project and the particular data collection must be filled out.

In case personal data needs to be transferred within the project, one of the following data sharing agreement will be put in place

– **Data Sharing Agreement**: agreement explaining the rights of two or more Parties that are using the data each one for their own purposes, separately.
– **Data Processing Agreement:** agreement explaining the rights of two (or more) Parties in which one of the Parties is the Data Controller and the other Party is Data

Processor – when one Party is using the data received by another one and it is using them only to help the first one to reach its own goal, like a service provider.
- **Joint Controller agreement**: agreement that explains the rights of Parties, two or more, that are responsible and decide together on the purposes and means of the data.

3.2 Review of the Documents

This manual and accompanying templates will be mandatory used during all experiments. The project consultant who is assigned to an experiment will support the respective SME filling in these documents. In addition, the ethics project lead coordinates the distribution of documents and is available for content and process questions.

All the above filled in documents will be collected in the beginning of the project and reviewed by a selection of members of the ethical data privacy committee and communicated back to the SMEs and consultant. The ethics checklist needs to be updated during the project and at the end of the project, at least before implementation. These updates do not need to be reviewed by the committee. Instead, updates should be discussed regularly during project meetings between SMEs and consultants.

3.3 Training

A training module has been made with the fundamentals of the Ethics manual content. The target audience are AIRISE consultants and SMEs who are in the validation and pilot-experiments. During the project, case studies will be collected to serve as training examples.

4 Discussion

For this work, the basis of the ethics manual are existing legislation (GDPR), and upcoming legislation (AI-act) including ethics (ALTAI) principles and principles for data management (FAIR). The main challenge lies in the implementation of the manual, making the principles understandable and the templates easy to fill in. Therefore, we kept the information in the ethics manual at a basic level and provided templates in a checklist form, with space for SME to comment and take notes. Furthermore, we provide continuous practical training and SMEs have the opportunity to liaise with their consultant and the ethics project lead. It is important that filling in the ethics template, will be repeated during the project and at the end, before implementation of the AI-solution. Ethics, data management and privacy are meant to be assessed in a life-cycle approach and is aimed to make the project better by creating awareness of the technology and safety aspects. This will ultimately improve the methodology of the AI-system.

Another point of attention is the novelty of responsible research practices amongst the project partners. Most of the organizations in the consortium are Research Technology Organization (RTO). Research ethics in general is strongly associated with human-related studies. RTOs have less extensive experience in research with humans. However, the ALTAI principles and the AI-act will be implemented horizontally, this includes the field of manufacturing. Although we expect the ethical principle: diversity, non-discrimination and fairness to be less applicable in manufacturing as compared to a medical AI case, a principle as a human – in the loop could be of high importance. The man-machine interaction will likely come under new scrutiny in light of AI-system implementation. In addition, some of the ethical principles such as robustness and transparency are of technical nature. It turned out very beneficial to have technical knowledge within the project consortium. This indicates that organizations who want to form ethics teams could benefit also from interdisciplinarity between AI ethics, technical experts. Due to the nature of these experiments, we do not expect highly sensitive personal data, nevertheless we wish to foresee in these scenarios by including privacy templates in the manual and training. Finally, SMEs who follow the FAIR principles will be able to properly manage their data for multiple purposes.

The manual will be used in 75 experiments in the upcoming years. Further research will primarily focus on the assessment of service delivery. Iteratively, the material will be further improved, ultimately a digital tool could make the process more efficient. Second of all, we will do content analyses on the ALTAI principles to contribute to solutions that benefit SMEs in manufacturing. This work could lay the foundation for new regulatory and legal frameworks, specific for the manufacturing field.

References

1. European Commission. A European strategy for data. A communication from the commission to the European parliament to the council, the European economic and social committee and the committee of the regions. Comm (2020)
2. European Commission, EU Grants How to Complete Your Horizon Ethics Self-Assessment (2) (2021)
3. EU Artificial Intelligence Act. https://artificialintelligenceact.eu/developments/. Accessed 31 May 2024
4. Independent high-level expert group on Artificial Intelligence, The assessment list for trustworthy Artificial Intelligence (ALTAI) for self-assessment (2020)
5. Openaire Homepage. https://www.openaire.eu/how-to-comply-with-horizon-europe-man date-for-rdm. Accessed 30 May 2024
6. Brintrup, A., Baryannis, G., Tiwari, A., Ratchev, S., Martinez-Arellano, G., Singh, J.: Trustworthy, responsible, ethical AI in manufacturing and supply chains: synthesis and emerging research questions. arXiv:2305.11581
7. European Commission. A Proposal for a Regulation of the European Parliament and of the Council Laying Down Harmonised Rules on Artificial Intelligence (Artificial Intelligence Act) and Amending Certain Union Legislative Acts, Article 17 (2021)
8. Commission implanting decision on a standardisation request to the European Committee for Standardisation (CEN) and the European Committee for Electrotechnical Standardisation (CENELEC) in support of safe and trustworthy artificial intelligence (2022)
9. AIRISE Homepage. https://airise.eu/. Accessed 31 May 2024

Author Index

© The Editor(s) (if applicable) and The Author(s) 2025
K. Alexopoulos et al. (Eds.): ESAIM 2024, LNME, pp. 287–288, 2025.
https://doi.org/10.1007/978-3-031-86489-6